Library and Information Services for Bioinformatics Education and Research

Shri Ram
Thapar University, India

A volume in the Advances in Library and Information Science (ALIS) Book Series

IGI Global
DISSEMINATOR OF KNOWLEDGE
www.igi-global.com

Published in the United States of America by
IGI Global
Information Science Reference (an imprint of IGI Global)
701 E. Chocolate Avenue
Hershey PA 17033
Tel: 717-533-8845
Fax: 717-533-8661
E-mail: cust@igi-global.com
Web site: http://www.igi-global.com

Library of Congress Cataloging-in-Publication Data

Names: Shri Ram, editor.
Title: Library and information services for bioinformatics education and research / Shri Ram, editor.
Description: Hershey, PA : Information Science Reference, [2017] | Series: Advances in library and information science | Includes bibliographical references and index.
Identifiers: LCCN 2016046933| ISBN 9781522518716 (hardcover) | ISBN 9781522518723 (ebook)
Subjects: LCSH: Medical librarianship. | Life sciences libraries--Collection development. | Bioinformatics--Information resources. | Libraries--Special collections--Bioinformatics.
Classification: LCC Z675.M4 L495 2017 | DDC 026/.61--dc23 LC record available at https://lccn.loc.gov/2016046933

This book is published in the IGI Global book series Advances in Library and Information Science (ALIS) (ISSN: 2326-4136; eISSN: 2326-4144)

British Cataloguing in Publication Data
A Cataloguing in Publication record for this book is available from the British Library.

Advances in Library and Information Science (ALIS) Book Series

ISSN:2326-4136
EISSN:2326-4144

Mission

The **Advances in Library and Information Science (ALIS) Book Series** is comprised of high quality, research-oriented publications on the continuing developments and trends affecting the public, school, and academic fields, as well as specialized libraries and librarians globally. These discussions on professional and organizational considerations in library and information resource development and management assist in showcasing the latest methodologies and tools in the field.

The **ALIS Book Series** aims to expand the body of library science literature by covering a wide range of topics affecting the profession and field at large. The series also seeks to provide readers with an essential resource for uncovering the latest research in library and information science management, development, and technologies.

Coverage

- Digitization Centers
- University Libraries in Developing Countries
- Storage Facilities
- Public Library Funding
- Diversity in Libraries
- State Library Agencies
- Collection Development
- Outsourcing of Library Services
- Library Buildings and Design
- Human Resources Management

The Advances in Library and Information Science (ALIS) Book Series (ISSN 2326-4136) is published by IGI Global, 701 E. Chocolate Avenue, Hershey, PA 17033-1240, USA, www.igi-global.com. This series is composed of titles available for purchase individually; each title is edited to be contextually exclusive from any other title within the series. For pricing and ordering information please visit http://www.igi-global.com/book-series/advances-library-information-science/73002. Postmaster: Send all address changes to above address.

Titles in this Series

For a list of additional titles in this series, please visit: www.igi-global.com

Academic Library Development and Administration in China
Lian Ruan (Illinois Fire Service Institute at the University of Illinois at Urbana-Champaign, USA) Qiang Zhu (Peking University, China) and Ying Ye (Nanjing University, China)
Information Science Reference • copyright 2017 • 391pp • H/C (ISBN: 9781522505501) • US $195.00 (our price)

Handbook of Research on Emerging Technologies for Digital Preservation and Information Modeling
Alfonso Ippolito (Sapienza University of Rome, Italy) and Michela Cigola (University of Cassino and South Latium, Italy)
Information Science Reference • copyright 2017 • 649pp • H/C (ISBN: 9781522506805) • US $275.00 (our price)

Information Seeking Behavior and Challenges in Digital Libraries
Adeyinka Tella (University of Ilorin, Nigeria)
Information Science Reference • copyright 2016 • 359pp • H/C (ISBN: 9781522502968) • US $185.00 (our price)

E-Discovery Tools and Applications in Modern Libraries
Egbert de Smet (University of Antwerp, Belgium) and Sangeeta Dhamdhere (Modern College of Arts, Science and Commerce, India)
Information Science Reference • copyright 2016 • 401pp • H/C (ISBN: 9781522504740) • US $195.00 (our price)

Technology-Centered Academic Library Partnerships and Collaborations
Brian Doherty (New College of Florida, USA)
Information Science Reference • copyright 2016 • 309pp • H/C (ISBN: 9781522503231) • US $165.00 (our price)

Space and Organizational Considerations in Academic Library Partnerships and Collaborations
Brian Doherty (New College of Florida, USA)
Information Science Reference • copyright 2016 • 367pp • H/C (ISBN: 9781522503262) • US $200.00 (our price)

IGI GLOBAL
DISSEMINATOR OF KNOWLEDGE

www.igi-global.com

701 E. Chocolate Ave., Hershey, PA 17033
Order online at www.igi-global.com or call 717-533-8845 x100
To place a standing order for titles released in this series,
contact: cust@igi-global.com
Mon-Fri 8:00 am - 5:00 pm (est) or fax 24 hours a day 717-533-8661

Editorial Advisory Board

Table of Contents

Detailed Table of Contents

the methodology for integration of information resources at a single platform. The integration platform is proposed shall highlight the role of the library in understanding the current best practices to deliver effective information to bioinformatics community. It will discuss the close relationship between data and information playing an extensive role in generation of bioinformatics knowledge. Further, a model has been proposed for the resource integration in the area of bioinformatics in order to provide a comprehensive platform for knowledge dissemination.

Information plays a vital role in bioinformatics to achieve the existing bioinformatics information technologies and to identify the needs of bioinformatics researchers. The most revolutionary development for bioinformatics resources is access to the internet because internet is pervasive in all bioinformatics work. Users required various sources of information for conducting bioinformatics research. The success of the information service is more likely to be achieved by adjusting the services to meet the specific needs of an individual.

Various biological databases are available online, which are classified based on various criteria for ease of access and use. All such bioinformatics database resources have been discussed in brief in this book chapter. The major focus is on most commonly used biological/bioinformatics databases. The authors provide an overview of the information provided and analysis done by each database, information retrieval system and formats available, along with utility of the database to its users. Most widely used databases have been covered in detail so as to enhance readers' understanding. This chapter will serve as a guide to those who are new to the field of bioinformatics database resources, or wish to have consolidated information on various bioinformatics databases available.

The largest digital repository of information, the World Wide Web keeps growing exponentially and calls for data mining services to provide tailored web experiences. This chapter discusses the overview of information retrieval, knowledge discovery and data mining. It reviews the different stages of data mining and introduces the wide spread biological databanks, their explosion, integration, data warehousing, information retrieval, text mining, text repositories for biological research publications, domain specific search engines, web mining, biological networks and visualization, ontology and systems biology. This chapter also illustrates some technical jargon with picture analogy for a novice learner to understand the concepts clearly.

Biomedical science is one filed where huge amount of information is generated, distributed over the internet and a number of software tools are also developed to generate information. The quantum of biomedical data along with the proliferation of new data integration technologies have made it important to adopt smart and fast network tools to access information in bioinformatics. It is important to make researchers in biomedical science aware of systematic approaches to access these information. One avenue to implement this approach is to make the biomedical information available through mobile technology which is still missing. It is heartening to see that there are some mobile initiatives taking place in biomedical sciences which provide handy tools for bioinformatics information seekers to access information. This paper is a review of such tools which will aid the library and information professionals to create information literacy in this field in future.

The Internet, which is an information super high way, has practically compressed the world into a cyber colony through various networks and other Internets. The development of the Internet and the emergence of the World Wide Web (WWW) as common vehicle for communication and instantaneous access to search engines and databases. Search Engine is designed to facilitate search for information on the WWW. Search Engines are essentially the tools that help in finding required information on the web quickly in an organized manner. Different search engines do the same job in different ways thus giving different results for the same query. Search Strategies are the new trend on the Web.

Purpose: The purpose of the study is to identify the library services and facilities provided to the Medical Scientists of JNIMS, Porompat. The study is limited to Medical Scientist of JNIMS who employed at twenty two (22) different medical departments of JNIMS. Design/Methodology/Approach: The study is based on survey method; questionnaire and interview method is used for collection of primary data. Hundred (100) questionnaires were distributed to the medical scientist of JNIMS. Findings: Services and facilities provided by the library are not satisfied by the medical scientist; library lack qualified manpower to function the library. Originality/Value: The study is part of the dissertation submitted to the Department of Library and Information Science, Manipur University for the year2014-2015. Article Type: Case Study

A survey was carried out of the information landscape within the students of Computer Science, Biology and Mathematics in the University of Swaziland which examined the research problems, important sources of information, the methods of access, information needs and seeking behavior of the users their assessment and the role of the Libraries since Librarian have to identify the information needs, uses and problems faced to meet the needs and requirement of the user. A total of 200 questionnaire were distributed. The survey indicated that majority of the students believe that the online resources play a very important role for their research and show positive attitude toward future bioinformatics usage and training. The study concluded that the training preferences of students need to be further explored.

A consortium is an association of independent libraries and/or library systems established by formal agreement, usually for the purpose of resource-sharing. The needs of special libraries cannot be fully addressed by regional organization because of its wide scope, thus, a consortium for specific group is deemed useful. This book chapter aims to describe the development of a local consortium and its role in building a culture of research, creating dynamic leadership and discussing how resource-sharing scheme goes beyond traditional inter-library loan. Using focus group discussion, the consortium members thresh out issues and concerns where collaborative research, dynamic leadership and resource- sharing pave way to enhance access to medical information.

With increase of information sharing over the internet or intranet, we require techniques to increase the availability of shared resource over large number of users trying to access the resources at the same time. Many techniques are being proposed to make access easy and more secure in distributed environment. Information retrieval plays an important to serve the most reliant data in least waiting, this chapter discuses all such techniques for information retrieval and sharing over the cloud infrastructure. Cloud Computing services provide better performance in terms of resource sharing and resource access with high reliability and scalability under high load.

Bioinformatics is rapidly growing, interdisciplinary field of science, where methods from information technology, computer science, mathematics, and statistics are used to solve problems of biological science. To access latest scholarly articles in such an important branch one cannot deny the importance of open access journals. In this chapter an attempt has been made to access the current status of open access journals of bioinformatics which are covered by Directory of Open Access Journals (DOAJ) on various parameters like country and language of publication, their currency, impact factor, article processing charges, copyright licensing model they are using, platform for hosting and their coverage in abstracting/indexing databases.

Foreword

Bioinformatics is an interdisciplinary field that develops and applies computational methods to analyze large collections of biological data; deploying computer technology to make sense of very large sets of data. There are huge amounts of data (big data) are generated by increasingly automated measuring devices. These data sets need to be stored, organized and analyzed to extract new insights and knowledge. In other words, "big data" needs to be converted into "smart data" and added to the knowledge base. Since bioinformatics involves organizing and analyzing large sets of data, and developing algorithms and statistical approaches to analyze and understand these data, it heavily relies on mathematical and statistical models and methodologies, as well as on computational tools and applications. It is complex field of academic activity. What distinguishes bioinformatics from other approaches, however, is its focus on developing and applying computational techniques to achieve this goal. Computing indeed has become a central component of scientific research in this 21st century.

Bioinformatics is now well established as a discipline with its own vocabularies, professional journals, and training programs. Many institutions provide bioinformatics support services through the Information Systems department because researchers' perception is that librarians are not capable of teaching non-bibliographic databases.

Dr. Shri Ram has taken on the laudable task editing a book, "Library and Information Services for Bioinformatics Education and Research". He is well qualified and has hands on recent experience in this field as evidenced by his doctoral thesis, He has persuaded his colleagues, who are directly involved in supporting bioinformatics research, to contribute to this volume. I am sure that the content provided by various authors has originated from their real time experience of working with bioinformatics community. This book will be a resource, not only for those working in the bioinformatics but also for the information professionals, who support bioinformatics education and research.

Gurdish Sandhu
University of East London – Dockland, UK

Preface

Bioinformatics involves the use of information science and technology to manage biological data and to support computer based experimentation by researchers. Biology has been described as evolving into an information-oriented science. Biomedical research now takes place in the context of an exploding information environment. Support for molecular biology researchers has been limited to traditional library resources and services in most academic health sciences libraries. But due to exploding information environment, the libraries have changed into information hub. Library services have been modified into the user centered services which involve the searching, consolidation, and integration to support research.

The book titled *Library and Information Services for Bioinformatics Education and Research* has been conceptualized based on the changing nature of the information requirement by the community. Everyday new information source and services are generated and being utilized by the bioinformatics community. Due to dynamic nature of information in bioinformatics, information scientists and libraries are always looking for new methodology to deliver resources at the desktop of the user. Various best practices and mechanism are being adopted to meet the expectation of the bioinformatics community. This books intended to gather those best practices and case study and share with others to exemplify these cases. The book has been divided into the fifteen chapters under single subject area "Library and Information Services for Bioinformatics Education and Research".

The chapter starts with changing nature of biomedical librarianship in post genomic era. The post genomic era starts after the completion of Human Genome Project (HGP), which has revolutionized the information in the subject of bioinformatics. As the information grown after HGP, the tremendous growth of data and information has taken place. It is very pertinent to organize that information in such a way that the information can be available in integrated manner. The role of librarian has become very crucial from identification of resources, consolidation and making them available to end user in helpful manner. The Second chapter deals with the establishing synergy between data and information through library services. This chapter deals with the development of comprehensive platform for delivery of information. The tool like *i*BIRA is quite helpful in this matter. The third chapter deals with the understanding the information needs of the bioinformatics community. It is very pertinent in the dynamic information environment, where there is a continued dynamism in information environment.

Chapter four is very important for those who are looking for information in the area of bioinformatics. The development of databases is one of the primary activities in the bioinformatics. The chapter outlines the availability of various databases in the bioinformatics. Data mining concept has now gained momentum in the data intensive field. Digging useful information from heap of information, different tools and techniques are being utilized, and data mining is playing a key role to mine the useful information. The concept of data mining techniques and its use has been discussed in chapter five. Chapter six deals with biological networks for data analysis, whereas chapter seven discusses the data mining activities in bioinformatics.

Mobile technology is advancing day-by-day so as the access to information through mobile devices. Chapter eight nicely elaborated the importance of mobile applications (Mobile Apps) in accessing biomedical information. Looking for right information through right techniques is very crucial. In this matter, chapter seven has put focus on the search strategies for digging bioinformatics information. The chapter highlighted some of the key issues on search strategies. Chapter ten is a case study about information seeking behavior of the users at a medical college library. Another case study on information seeking behavior has been discussed for University of Swaziland in chapter eleven.

Leadership in one of the factor which impact overall growth of individual as well as institution, in fact, the libraries leadership initiatives help student and research scholar in terms of new services and resources. The leadership initiatives like consortia have actually helped a lot in terms of optimizing the user of resources and managing library budget. Chapter twelve has discussed the leadership initiatives and role of local library consortia. Chapter thirteen deals which cloud services in accessing bioinformatics information while chapter fourteen is a good resource for researcher for choosing journals which are published in open access mode. The final chapter deals with the availability of web based information resources on *Zea mays* which is third most used food grain in the world.

Each author has put forward their effort to organize the chapter in nice way and I am sure that the target audience certainly benefitted with the content embodied.

Acknowledgment

The book titled *Library and Information Services for Bioinformatics Education and Research* is dedicated to my parents and I am blessed with their kind blessings. I would like to express my sincere thanks to all the authors who have accepted my call and contributed their papers. I would like to thank all the reviewers who have spared their valuable time to review the papers and provided critical remarks on each paper. My sincere thanks goes to my editorial advisors who have suggested my various key aspects for editing a book and guided me to bring out this book. I would like to express my sincere thanks to IGI Global who have accepted my proposal and approved the book for the benefit of the bioinformatics community. The entire team at IGI-Global needs a special thanks from the bottom of heart. At the end, I would like to thank my wife and two children who have helped and motivated me to undertake this assignment and always stood along with me.

Shri Ram
Thapar University, India

Acknowledgment

Chapter 1
Biomedical Librarianship in the Post-Genomic Era

Shubhada Prashant Nagarkar
Savitribai Phule Pune University, India

ABSTRACT

Post genomic era is known for the explosive growth in biomedical information. Bibliographic and sequence databases are increasing continuously and have voluminous data sets. Biomedical librarians are facing challenges in retrieval of relevant information from these electronic databases and related sources of information. This chapter discusses the changing role of biomedical librarians in post genomic era. The chapter covers features of the biomedical librarianship including library collection development, users' information needs and strategies adopted to provide services. Moreover, it focuses on the competencies required by librarians to face the challenges of management of information and services needed by biomedical researchers in the post genomic era.

INTRODUCTION

The librarianship is a trinity of acquisition, organization and dissemination of information, in which acquisition relates to the proper selection of library materials, organization to their preparation for efficient/effective use, and dissemination to the processes of making the contents available to the users. To achieve this, librarians should act as a mediator and must know the intellectual contents of information resources, user information needs and the methods to bring both contents and users

DOI: 10.4018/978-1-5225-1871-6.ch001

together (Shera, 1972). Trinity of librarianship has become challenging over the period due to remarkable changes in the containers of information, interdisciplinary approach of researchers and new communication technology for dissemination of information. Containers of information changed over the period i.e. from paper to electronic. Organization of information has become more challenging to meet the interdisciplinary information needs of users.

The present chapter discusses the trinity of librarianship in the field of Bioinformatics and the changing role of librarian in bringing information and user together. This particular subject area is selected due to two reasons. Firstly, the author has worked as a practicing librarian in the field of Bioinformatics for almost seventeen years and secondly, the librarianship in this field has changed to a great extent especially in the post-genomic era. The chapter has three main sections viz. library collection, user information needs and library services to Bioinformaticians. Each of this section discusses the challenges for librarians and efforts by librarians. The initial part of the chapter talks about the information explosion in the field of biology, interdisciplinary nature of research and features of post genomic librarianship. The chapter has the context of Bioinformatics libraries and services evolved during last 10 years which cater to the needs of Bioinformatics community including students, teachers and researchers.

Bioinformatics in the Post Genomic Era

There is a remarkable difference in Pre and Post genomic era in respect to biological research and information. The research in pre genomic era was non interdisciplinary and individual scientist worked for individual research. In the post genomic era research has become interdisciplinary and there is integration between parallel studies in the allied research fields (Torshin, 2006). The pre genomic era or the beginning of the Bioinformatics era is marked by the contributions during the 1960s by Late Dr. Margaret Dayhoff for the compilation and analysis of large data sets of protein sequences to study the molecular evolution (Dayhoff, 1965). The fusion of biology and computer—Bioinformatics—evolved in the late 20th century. The major event in computing, the introduction to the World Wide Web in 1990 which was coincided with the beginning of the Human Genome Project (HGP) (Bergeron, 2003). WWW and Human Genome Project significantly represents the convergence of computing, communication, and molecular biology (Benoît, 2006; Bergeron, 2003). The HGP is an example where researchers from a variety of disciplines viz. biotechnology and molecular biology, computer science, engineering, physics, chemistry, mathematics, statistics and medicine worked together to reveal all (approximately 21,000 genes) human genes. The collaboration among scientists is required in managing the large quantities of data necessary to reveal biological relationships and in using innovative techniques to locate, aggregate, manipulate, and present such data through user-friendly, cross-platform applications (Goodman, 2003; MacMullen & Denn, 2005). Apart from HGP several other projects including research into specific organism or cell type, gene expression, metabolic pathways, regulatory networks, and protein-protein interaction data, have been responsible for explosive growth in the generation of biological information (Benoît, 2006). This explosive growth leads towards the creation of various databases, databanks, and software tools. The 2015 *Nucleic Acids Research* Database Issue contains 172 papers that include descriptions of 56 new molecular biology databases, and updates on existing 115 databases (Galperin et. al, 2014).

As the discipline evolved and its scope became broader, the demand for trained human resource started growing. This necessitated the establishment of formal training programmes. Very few research institutes started short terms training programmes in Bioinformatics which later grown into diploma, masters, and Ph.D. programmes (Magana et. al., 2014; Kulkarni-Kale, Sawant, & Chavan, 2010).

The tremendous growth in published biological literature is the mark of the post genomic era. In literature database, the term Bioinformatics appeared in the year 1997 with the description "A field of biology concerned with the development of techniques for the collection and manipulation of biological data, and the use of such data to make biological discoveries or predictions. This field encompasses all computational methods and theories for solving biological problems including manipulation of models and datasets" (http://www.ncbi.nlm.nih.gov/mesh/?term=bioinformatics). PubMed search results on "bioinformatics / computational biology" as a major MeSH term shows 63224 (retrieved on 15/2/2016) research publications which are growing geometrically.

It has also been noticed that new key terms related to Bioinformatics appeared in the literature viz. computational molecular biology, computational biology, computational biosciences, computational biomedicine, biomedical informatics, and biological information to name a few (Kangueane, 2009)

As discussed above, Bioinformatics is interdisciplinary in nature and researchers from "Information Science" also contributed to this new discipline. The tools and techniques of "organization of information" are highly applicable to this discipline. The librarian playing a key role in organizing bioinformatics using different methodologies and one such methodology was reported on the integration of various types of bioinformatics information resources – *i*BIRA (Ram & Rao, 2012).

The next part of the chapter discusses the impact of post genomic era on libraries especially biology libraries. It has affected the collection development, information needs of users, and information services. It also lists the challenges for librarians under each section. The chapter concludes with the core competencies needed by the Bioinformatics librarians.

COLLECTION DEVELOPMENT IN BIOINFORMATICS

Availability and accessibility of information are different in a pre and post genomic era. Collection development decisions are largely influenced due to two issues viz. information overload and interdisciplinary nature of research. As discussed above, the research in the pre genomic era was not interdisciplinary and hence the publications were limited. Till date very little research has been conducted on developing library collection in Bioinformatics (Martin, 2013) Collection development in Bioinformatics libraries need liaison librarian having knowledge of core subject and knowing the information needs and information seeking behavior pattern of users which will assist in building the balanced and meaningful collection in libraries (Martin, 2013). A specific subject focus of a collection is dependent on the particular academic program and research area. Due to the interdisciplinary nature of Bioinformatics, the collection should include publications from other related areas. Following are some of the criteria to be considered while developing the Bioinformatics library collection.

Books

- Book collection should be from various related disciplines and should support teaching learning and research.
- Faculty and student suggestions should be taken into consideration.
- Use of hard and e-copies of publisher's catalogue should be made habitual.
- E-mail alerts of well-known book publishers and suppliers' web sites for the notifications of forthcoming books and other publications should be checked.

Journals, E-Journals and Online Databases

- Subscribe to core and allied journals having high impact factor and published by renowned publishers and indexed in databases like Web of Science and Scopus.
- Provide access to electronic journals and e-books and other multimedia resources like video talks for students.
- Subscribe to E-resources on the basis of evaluation criteria like subject areas covered, status of publisher, access during and after subscription period, user interfaces, remote access to users, preservation and future access, etc.
- Awareness of Open Access journals in the field.
- Be aware of publishers' copyright and self-archiving policies from projects like SHERPA/RoMEO and plagiarism policies of each publisher.

- Knowledge of availability of multimedia resources like video talks, lecture series, etc.
- Knowledge of Institutional Repositories including electronic thesis and dissertations.
- Use of Research Information Network sites like ResearchGate, Academia.edu, Faculty of 1000, etc.

A list of core Bioinformatics journals is provided in the Appendix in Table 2. Apart from these core journals there are many more journals like *Journal of Molecular Biology, Nucleic Acids Research, Protein Science,* and *Journal of Computational Biology* supporting Bioinformatics-related work. Based on the ISI Web of Science Journal Citation Report (JCR 2015), a ranking of journals is presented in Table 1.

Two journals, *Genome Research* and *Genome Biology* have the highest impact factor (JCR, 2015) with 11.351 and 11.313 respectively. These two journals have shown continued growth in its impact factor. The status of impact factor for top nine journals with the highest impact factor is given in Figure 1.

The impact factor is one of the parameters used for publishing research articles in a given field of study (Garfield 2003; Kaltenborn & Kuhn 2004). The bioinformatics scientists and research would also be considering the impact factor as one of the criteria for publication of the research work. Certain challenges arise while selecting right journals for publishing and these may be related to:

- Finding out high quality core and interdisciplinary journals, books and e-resources.
- Help of subject experts to understand the subject and the interdisciplinary.
- Finance and budget.
    - Generally the Bioinformatics journals have high subscription charges and are available in both print and electronic format.
- Electronic Resources Management.
- Classification of interdisciplinary books.
- Understanding the citation analysis methods for building good collection of journals.

There are various tools available which help in understanding the ranking of journals from a particular field of research. InCite from *Web of Science*, Journal Citation Report from *Web of Science*, SciVal from *Elsevier*, SciMago Journal Ranking from *SCOPUS* are the available tools which can be helpful in choosing right journals for probable publication.

Table 1. Key journals in bioinformatics sorted by Impact Factor

Name of Journal	Rank	Category Normalized Citation Impact	Journal Normalized Citation Impact	Immediacy Index	Eigenfactor	5 Year Impact Factor	Journal Impact Factor
Genome Research	8	0.72	0.22	2.322	0.12414	14.381	11.351
Genome Biology	5	1.37	0.52	1.826	0.087	13.168	11.313
Nucleic Acids Research	1	5.86	2	2.576	0.36513	8.647	9.202
Briefings in Bioinformatics	10	2.7	1.07	1.337	0.0141	6.778	8.399
Molecular & Cellular Proteomics	11	1.96	0.9	1.356	0.05272	6.632	5.912
Bioinformatics	2	11.01	4.5	1.167	0.1851	7.685	5.766
PLoS Computational Biology	13	1.55	1.34	0.568	0.08546	5.017	4.587
Journal Of Proteome Research	6	1.45	1.01	1.019	0.05243	4.341	4.173
Proteomics	4	1.47	1.29	0.858	0.02753	3.666	4.079
BMC Genomics	7	0.69	0.82	0.52	0.09426	4.278	3.867
Biochimica Et Biophysica Acta-Proteins and Proteomics	22	0.68	1.15	0.783	0.01942	3.28	3.016
Database-the Journal of Biological Databases and Curation	23	0.46	0.68	0.473	0.00931	3.983	2.627
Molecular Genetics and Genomics	19	0.68	0.71	0.725	0.00448	2.858	2.622
Proteins-Structure Function and Bioinformatics	9	0.4	0.55	0.643	0.02256	2.725	2.499
BMC Bioinformatics	3	0.82	1.44	0.396	0.06818	3.443	2.435
Genomics	14	1.52	2.02	0.556	0.00798	2.896	2.386
Gene	12	1.03	1.49	0.525	0.02957	2.258	2.319
BMC Systems Biology	16	1.23	1.73	0.388	0.01665	2.9	2.213
IEEE-ACM Transactions on Computational Biology and Bioinformatics	15	0.35	0.75	0.203	0.00583	1.778	1.609
Evolutionary Bioinformatics	21	0.47	1.06	0.167	0.00133	1.58	1.404
Journal of Bioinformatics and Computational Biology	24	0.35	0.48	0.564	0.00168	n/a	0.785
Current Bioinformatics	17	0.07	0.68	0.016	8.30E-04	0.93	0.77
International Journal of Data Mining and Bioinformatics	18	0	0	0.068	3.40E-04	0.717	0.528

Figure 1. Top journals in bioinformatics and impact factor trends

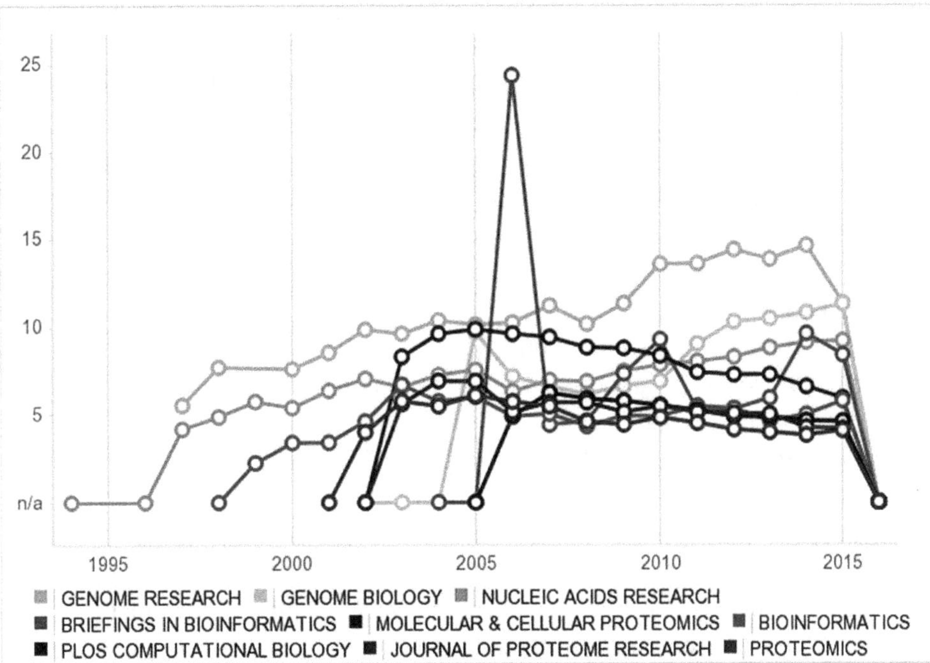

**For a more accurate representation of this figure, please see the electronic version.*

INFORMATION NEEDS OF BIOINFORMATICIANS

Understanding the user information needs would help in planning the collection development and to provide information services to users. There are very few studies focusing on information needs of bioinformaticians. As the field is multidisciplinary, users need information from variety of resources viz. literature databases, protein and sequence databanks, genome databases, etc. They need information about particular genes, proteins, experimental techniques and biological processes and they want access to non-clinical protocols, and information on grants and educational opportunities. Due to the availability of several Bioinformatics analysis tools, researchers need even more guidance as to which resources to use and how to use them (Shipman, et.al, 2005). US National Institutes of Health (NIH) conducted an information needs study of clinical specialists and biomedical researchers to inform library services and contribute to a broader understanding of information use in academic and research settings (Grefsheim & Rankin, 2007).

Cunningham conducted a survey of health science libraries and how they are catering to the information needs of biotechnologists working in nine medical center. Results indicated that librarians use MEDLINE (now PubMed) database to answer the queries of researchers. The survey of both librarians and biotechnologists indicated that they need special training to serve the purpose. The "Biotechnology Awareness Study Part 1 and 2 were successful and both became aware of biotechnology resources (Cleveland, et.al. 2007; Cunningham, et.al. 1991).

University of Florida's Health Science Center Libraries (HSCL) libraries conducted surveys of use of mobile devices by students to understand current use, preferences, and future needs for smartphone support in an academic health science environment. Bushhousen and others reviewed the variety of uses of the smart phone by biomedical students for education and research purposes viz. scientific graphing, examining 3D images of chemical compounds, viewing video tutorials etc. (Bushhousen et. al., 2013).

Above studies indicate that to understand the user information needs librarians need subject knowledge and users need training in retrieving the information from literature as well as other databases. Moreover, they need to keep studying the information needs of biomedical researchers in changing environment.

Challenges

- Need to understand the ever changing interdisciplinary information needs which requires continuous research.
- Work like embedded librarian to know the requirements of researchers.
- To design Information Literacy courses to enhance the Information Search skills of users.

LIBRARY SERVICES

In the post genomic era it is observed that Bioinformatics libraries are providing variety of services along with traditional services to Bioinformatics researchers. New services include references services through electronic devices, designing new training and teaching programs for faculty, and developing library portals.

Reference Services

Librarians can make a significant impact by employing online Current Awareness Services (CAS) and Selective Dissemination of Information (SDI) to keep up with the plethora of available resources (Osterbur, et. al. 2006; Alpi, 2003). Internet has changed over the time and has become interactive. It includes Web 2.0 tools viz. RSS feeds, email alerts, Google gadgets, wikis, blogs, online groups, news aggregators, social networking and social book marking tools, etc. With the help of these tools librarians can create, annotate, review, reuse and represent the information in new ways (Arndt & Currie, 2010; Hey, 2012). Librarians must take the advantage of these tools to save the time of the researchers (Nagarkar, 2011).

Librarians in recent years conducted various surveys to understand the use of modern devices by biomedical students (Mi, et al., 2016). This systematic review of 20 studies selected from PubMed published during 2010-2015. The results indicate that use of mobile devices is beneficial to students as well as they face challenges in using the same. There is significant variability across the mobile programs and resources accessed. The University at Buffalo Health Sciences Library purchased iPads to develop embedded reference and educational services. Results indicated that iPad can be used to meet the library user's needs outside of the physical library space (Stellrecht & Chiarella, 2015). Darcy, Text Reference Coordinator at SUNY Purchase College Library, addresses challenges and misconceptions surrounding SMS reference services and suggests best practices for SMS reference interactions and staffing. Librarians need to explore new techniques like "mobile reference" and return to long-standing practices like "saving the time of the user" and telephone ready reference (Darcy, 2014).

Role of Librarians

Further, librarians who are "Masters of the Info Universe" (Kerith, 2011) should recognize the complexity of the scientists' queries (Rein, 2006; Tennant, 2005) and design good search strategies for each and every database used. They need to keep constant watch on the new developments of search engines, reference management tools as well as services offered by database providers. The immediate need is to design personalized information services to users to save their time (Youngkin, 2010; Wu & Li, 2007) rather than the traditional reference services. Barrett mentions that the use of reference services is declining while the inception of chat and email reference services broaden the patron base (Barrett, 2010). Librarians should take the advantage of Web 2.0 tools to promote the use of library services (Ivie et al., 2011).

Training Programs

The major problem is that scientists are not formally trained in "information retrieval" techniques. They often use the resources they know, get some of the information they need, and move on. As the number and complexity of available resources grow, users generally lose track of what they have done. They need guidance on using various tools under different conditions. Librarians' role under such a situation should be dynamic. Specially trained librarians could work as embedded librarians working with scientists and establishing their reputation as problem solvers (Konieczny, 2010; Rein, 2006; Hoffman & Ramin, 2010)

Many libraries started specialized programs for bioinformaticians. For these purpose, they initially hired Bioinformatics specialists having strong background in molecular biology to understand the campus wide Bioinformatics related information needs of researchers. Washington University's Bernard Becker Medical Library developed three courses viz. Sequence Similarity Search, Genetic Variation, and Human Genome Resources. These courses were offered twice per semester. Moreover, library scheduled in-depth consultations with faculty and researchers to address their specific needs. These consultations led to requests for other software packages to be purchased. (Yarfitz & Ketchell, 2000, Wang, et.al. 2007, Yi-Bu, et.al. 2007). Purdue University Library (PUL) having specialization in agriculture, engineering, biomedical and applied life sciences with an increasing focus on Bioinformatics, started a specialized "Bioinformatics Week" focused Bioinformatics instruction to launch system-wide, library-based Bioinformatics services. For the purpose they also hired molecular bioscience specialist to discover, engage, and support Bioinformatics needs across the campus (Rein, 2006). The specialist is also engaged in training PUL librarians in Bioinformatics to provide a sustaining culture of library-based Bioinformatics support and understanding of Purdue's Bioinformatics-related decision and policy making. Norris Medical Library established Bioinformatics Service Program in 2005. Library over the period assessed users' Bioinformatics needs, acquired additional funds, established and expanded service offerings, and explored additional roles in promoting on-campus collaboration. In 2013 it is noticed that the library-based Bioinformatics service programs can become a key part of an institution's comprehensive solution to researchers' ever-increasing Bioinformatics needs (Li, et.al. 2013).

In above mentioned programs, it is noticed that librarians worked with subject experts to know the subject and to understand their information needs. This opportunity enhanced their subject knowledge which they further used to assist users to search relevant information.

Library Portals

Librarians, made efforts to organize e-resources on web and provide access to users to save the time of user as well as librarian. Information portals came into being to bridge the gap between the rising information needs and the rapidly growing number of online Bioinformatics resources. The Health Sciences Library System (HSLS) at the University of Pittsburgh have created the Online Bioinformatics Resources Collection (OBRC). OBRC is aimed at becoming a one-stop guided information gateway to the major Bioinformatics databases and software tools on the Web. OBRC is available at the University of Pittsburgh's HSLS Web site (Yi-Bu et al., 2007). The portals are also designed to provide personalized information services to bioinformaician (Nagarkar, 2011). Portals created by University of Queensland, University of Minnesota, and John Hopkins Institute are some other examples.

Challenges

- Adopt and learn new Information and Communication Technologies (ICTs) and new mobile devices.
- Formulating effective Boolean queries which require domain expertise and knowledge of contents and facilities of the databases (Lacroix, 2002).
- Text mining and data mining techniques to mine the structured data and patterns.
- Development of plans for new services, recruitment and training of specialized staff, and establishment of collaborations with Bioinformatics centers and experts (Geer, 2006).

CONCLUSION

Core Competencies of Bioinformatics Librarians

The following skills and competencies will equip Bioinformatics librarian to face the challenges in collection development, understanding information needs of researchers and providing value added library services:

1. **Core Subject Knowledge:**
    a. Be familiar with latest trends in the fields by participating in training programmes / conferences and seminars.
    b. Participate in awareness workshops other than library and information science fields to understand the trends in research.
    c. Awareness of latest trends in publishing.
    d. Awareness of copyright and intellectual property laws.
2. **Information Organization Skills:**
    a. Classification and cataloguing of library collection.
    b. Use of control vocabulary tools like MeSH and other thesauruses.
    c. Development and design of library portals.
3. **Computing Skills:**
    a. Mobile and cloud based information services.
    b. Skills of effective use of various search engines.
    c. Tools for library automation, digital libraries and content management, data and text mining.

4. **Collaborative Skills:**
    a. In-depth knowledge of research and development programmes in the organization.
    b. Collaborative activities with other faculty on the campus or within the organization.
    c. Work like embedded librarian and become a team member of scientific activities.
5. **Communication Skills:**
    a. Negotiation and communication skills with publishers and suppliers.
    b. Communication skills with users to understand information needs.
6. **Information Literacy Skills:**
    a. Design and development of information literacy programmes.
    b. Teaching abilities.

SUGGESTION

To develop these competencies into future Bioinformatics librarians, library schools from developing countries should design special training courses to serve Bioinformatics community.

ACKNOWLEDGMENT

The author would like to thank Mrs. Asha Umarani, Retd. Associate Professor, Department of Library and Information Science, and Savitribai Phule, Pune University, Pune, India for her valuable guidance and help in editing the chapter.

REFERENCES

Alpi, K. (2003). Bioinformatics training by librarians and for librarians: developing the skills needed to support molecular biology and clinical genetics information instruction. *Issues in Science and Technology Librarianship, 37*(Spring). Retrieved from http://www.istl.org/03-spring/article1.html

Arndt, T., & Currie, J. P. (2010). Web 2.0 for reference services staff training and communication. *Reference Services Review, 38*(1), 152–157. doi:10.1108/00907321011020789

Barrett, F. A. (2010). *An analysis of reference services usage at a regional academic health sciences library*. Retrieved from https://indigo.uic.edu/handle/10027/7618

Benoît, G. (2006). Bioinformatics. *Annual Review of Information Science & Technology*, *39*(1), 179–218. doi:10.1002/aris.1440390112

Bergeron, B. P. (2003). *Bioinformatics Computing*. Prentice Hall Professional.

Bushhousen, E., Norton, H. F., Butson, L. C., Auten, B., Jesano, R., David, D., & Tennant, M. R. (2013). Smartphone uses at a university health science center. *Medical Reference Services Quarterly, 32*(1), 52–72. doi:10.1080/02763869.2013.749134

Cleveland, A. D., Hannigan, G. G., Bedard, M., Philbrick, J. L., & Turner, P. M. (2007). *Recruiting the next generation of biomedical sciences librarians: Meeting increasingly complex information needs by building on a biomedical sciences education foundation*. Retrieved from http://www.icml9.org/program/track9/public/documents/Ana%20D-110009.doc

Cunningham, D., Grefsheim, S., Simon, M., & Lansing, P. S. (1991). Biotechnology awareness study, Part 2: Meeting the information needs of biotechnologists. *Bulletin of the Medical Library Association, 79*(1), 45–52. Retrieved from http://www.ncbi.nlm.nih.gov/pmc/articles/PMC225484/ PMID:1998819

Dayhoff, M. O. (1965). *Atlas of protein sequence and structure*. Retrieved from http://agris.fao.org/agris-search/search.do?recordID=US201300600070

Garfield, E. (2003). The meaning of the Impact Factor. *International Journal of Clinical and Health Psychology*, *3*(2), 363–369.

Geer, R. C. (2006). Broad issues to consider for library involvement in bioinformatics. *Journal of the Medical Library Association: JMLA*, *94*(3), 286. Retrieved from http://www.ncbi.nlm.nih.gov/pmc/articles/PMC1525323/ PMID:16888662

Gervasio, D. I. (2014). Redefining Virtual: Leveraging Mobile Librarians for SMS Reference. *International Journal of Digital Library Systems*, *4*(2), 44–69. doi:10.4018/IJDLS.2014070104

Goodman, L. (2003). *Making a genesweep: it's official. In BioIT* (p. 12). World News.

Grefsheim, S. F., & Rankin, J. A. (2007). Information needs and information seeking in a biomedical research setting: A study of scientists and science administrators. *Journal of the Medical Library Association: JMLA*, *95*(4), 426–434. doi:10.3163/1536-5050.95.4.426 PMID:17971890

Hey, T. (2012). The Fourth Paradigm–Data-Intensive Scientific Discovery. In E-Science and Information Management. Springer. doi:10.1007/978-3-642-33299-9_1

Hoffman, S., & Ramin, L. (2010). Best practices for librarians embedded in online courses. *Public Services Quarterly*, *6*(2-3), 292–305. doi:10.1080/15228959.2010.497743

Ivie, T., McKay, B., May, F., Mitchell, J., Mortimer, H., & Walker, L. A. (2011). Marketing and promotion of library services using Web 2.0: An annotated mediagraphy. *The Idaho Librarian, 61*(1). Retrieved from http://works.bepress.com/lizzy_walker/3/

Kaltenborn, K. F., & Kuhn, K. (2004). The journal impact factor as a parameter for the evaluation of researchers and research. *Revista Espanola de Enfermedades Digestivas*, *96*(7), 460–476. doi:10.4321/S1130-01082004000700004

Kangueane, P. (2009). *Bioinformation Discovery: Data to Knowledge in Biology*. Berlin: Springer Science & Business Media. doi:10.1007/978-1-4419-0519-2

Kerith, P. M. (2011). *Librarians: Masters of the info universe*. Retrieved from http://edition.cnn.com/2011/LIVING/04/12/librarians.masters.of.universe/

Konieczny, A. (2010). Experiences as an Embedded Librarian in Online Courses. *Medical Reference Services Quarterly*, *29*(1), 47–57. doi:10.1080/02763860903485084 PMID:20391164

Kulkarni-Kale, U., Sawant, S., & Chavan, V. (2010). Bioinformatics education in India. *Briefings in Bioinformatics*, *11*(6), 616–625. doi:10.1093/bib/bbq027 PMID:20705754

Lacroix, Z. (2002). Biological data integration: wrapping data and tools. *Information Technology in Biomedicine, IEEE Transactions on*, *6*(2), 123–128. Retrieved from http://ieeexplore.ieee.org/xpls/abs_all.jsp?arnumber=1006299

Li, M., Chen, Y.-B., & Clintworth, W. A. (2013). Expanding roles in a library-based bioinformatics service program: A case study. *Journal of the Medical Library Association: JMLA*, *101*(4), 303–309. doi:10.3163/1536-5050.101.4.012 PMID:24163602

MacMullen, W. J., & Denn, S. O. (2005). Information problems in molecular biology and bioinformatics. *Journal of the American Society for Information Science and Technology*, *56*(5), 447–456. doi:10.1002/asi.20134

Magana, A. J., Taleyarkhan, M., Alvarado, D. R., Kane, M., Springer, J., & Clase, K. (2014). A survey of scholarly literature describing the field of bioinformatics education and bioinformatics educational research. *CBE Life Sciences Education, 13*(4), 607–623. doi:10.1187/cbe.13-10-0193 PMID:25452484

Martin, V. (2013). Developing a Library Collection in Bioinformatics. In Bioinformatics: Concepts, Methodologies, Tools, and Applications (pp. 130 - 150). Idea Group Inc. (IGI).

Mi, M., Wu, W., Qiu, M., Zhang, Y., Wu, L., & Li, J. (2016). Use of Mobile Devices to Access Resources Among Health Professions Students: A Systematic Review. *Medical Reference Services Quarterly, 35*(1), 64–82. doi:10.1080/02763869.2016.1117290 PMID:26794197

Nagarkar, S. (2011). *Web-based Reference Services to Bioinformaticians: Challenges for librarians*. Retrieved from http://conference.ifla.org/past/2011/111-nagarkar-en.pdf

Osterbur, D. L., Alpi, K., Canevari, C., & Corley, P. M. (2006). Vignettes: diverse library staff is offering diverse bioinformatics services. *Journal of the Medical Library Association, 94*(3), E188–91.

Ram, S., & Rao, N. L. (2012). iBIRA – integrated bioinformatics information resource access: Organizing the bioinformatics resourceome. *Reference Services Review, 40*(2), 326 – 343.

Rein, D. C. (2006). Developing library bioinformatics services in context: The Purdue University Libraries bioinformationist program. *Journal of the Medical Library Association: JMLA, 94*(3), 314. Retrieved from http://www.ncbi.nlm.nih.gov/pmc/articles/PMC1525331/ PMID:16888666

Shera, J. H. (1972). *The Foundations of Education for Librarianship*. New York: Becker and Hayes.

Shipman, J. P., Barbara Watstein, S., & Tennant, M. R. (2005). Bioinformatics Librarian: meeting the information needs of genetics and bioinformatics researchers. *Reference Services Review, 33*(1), 12–19. doi:10.1108/00907320410519333

Stellrecht, E., & Chiarella, D. (2015). Targeted Evolution of Embedded Librarian Services: Providing Mobile Reference and Instruction Services Using iPads. *Medical Reference Services Quarterly, 34*(4), 397–406. doi:10.1080/02763869.2015.1082372

Tennant, M. R. (2005). Bioinformatics Librarian: Meeting the information needs of genetics and bioinformatics researchers. *RSR. Reference Services Review*, *33*(1), 12–19. doi:10.1108/00907320410519333

Torshin, I. Y. (2006). *Bioinformatics in the Post-genomic Era: The Role of Biophysics*. New York: Nova Publishers.

Wang, L., Lipsey, K., Murray, C., Prendergast, N., & Schoening, P. (2007). The Bioinformatics Program at Washington Universitys Bernard Becker Medical Library: Making it happen. *Medical Reference Services Quarterly*, *26*(2), 87–98. doi:10.1300/J115v26n02_08 PMID:17522011

Wu, W. G., & Li, J. (2007). RSS made easy: A basic guide for librarians. *Medical Reference Services Quarterly*, *26*(1), 37–50. doi:10.1300/J115v26n01_04 PMID:17210548

Yarfitz, S., & Ketchell, D. S. (2000). A library-based bioinformatics services program. *Bulletin of the Medical Library Association*, *88*(1), 36–48. PMID:10658962

Yi-Bu, C., Chattopadhyay, A., Bergen, P., Gadd, C., & Tannery, N. (2007). The Online Bioinformatics Resources Collection at the University of Pittsburgh Health Sciences Library System--a one-stop gateway to online bioinformatics databases and software tools. *Nucleic Acids Research*, *35*(1), D780–D785. PMID:17108360

Youngkin, A. (2010). Librarian-controlled RSS: A novel approach to literature search follow-up. *Journal of Hospital Librarianship*, *10*(2), 123–131. doi:10.1080/15323261003680028

KEY TERMS AND DEFINITIONS

Bioinformatics Librarian: A person who dedicated towards the assessment of information, selection, consolidation and help bioinformatics community with the desired information.

Bioinformatics: A subject which encompassed the application of computer science and mathematics for the biological data analysis.

Genomics: The Science which deals with structure and function of the gene, the basic unit of life.

Information Services: A service which provides information.

APPENDIX

Table 2. List of core bioinformatics journals

Sr. No.	Journal	Country
1.	*Advances and Applications in Bioinformatics and Chemistry.*	New Zealand
2.	*Advances in Bioinformatics.*	Egypt
3.	*Applied Bioinformatics.*	New Zealand
4.	*Bioinformatics*	United Kingdom
5.	*Bioinformatics and Biology Insights.*	New Zealand
6.	*BMC Bioinformatics*	United Kingdom
7.	*Briefings in Bioinformatics*	United Kingdom
8.	*Computational systems bioinformatics / Life Sciences Society. Computational Systems bioinformatics Conference.*	United Kingdom
9.	*Current Bioinformatics*	Netherlands
10.	*Current Protocols in Bioinformatics*	United States
11.	*Eurasip Journal on Bioinformatics and Systems Biology*	United States
12.	*Evolutionary Bioinformatics*	New Zealand
13.	*Genomics Proteomics Bioinformatics*	China
14.	*IEEE/ACM Transactions on Computational Biology and Bioinformatics.*	United Kingdom
15.	*International Journal of Bioinformatics Research and Applications*	United Kingdom
16.	*International Journal of Data Mining and Bioinformatics*	United Kingdom
17.	*IPSJ Transactions on Bioinformatics. Japan.*	
18.	*Journal of Bioinformatics and Computational Biology*	Singapore
19.	*Journal of Clinical Bioinformatics*	United Kingdom
20.	*Journal of integrative bioinformatics*	Germany
21.	*Journal of Proteomics and Bioinformatics*	India
22.	*Mathematical Biology and Bioinformatics.*	Russian Federation
23.	*Open Bioinformatics Journal.*	Netherlands
24.	*Proceedings - 2004 IEEE Computational Systems Bioinformatics Conference, CSB 2004.*	Unite States.
25.	*Proceedings - Fourth IEEE Symposium on Bioinformatics and Bioengineering, BIBE 2004.*	United States
26.	*Proteins: Structure, Function, and Genetics*	The United States.
27.	*Trends in Bioinformatics.*	Pakistan

Chapter 2
Library Services for Bioinformatics:
Establishing Synergy Data Information and Knowledge

Shri Ram
Thapar University, India

ABSTRACT

Bioinformatics is an emerging data intensive discipline. The community and information resources and sources are heterogeneous. It is the role of library to provide a comprehensive platform to deliver effective information services to the community. The paper discusses the status of various bioinformatics information resources available for the community. It is essential to search, consolidate and made information resources available to the community. The paper also discusses the methodology for integration of information resources at a single platform. The integration platform is proposed shall highlight the role of the library in understanding the current best practices to deliver effective information to bioinformatics community. It will discuss the close relationship between data and information playing an extensive role in generation of bioinformatics knowledge. Further, a model has been proposed for the resource integration in the area of bioinformatics in order to provide a comprehensive platform for knowledge dissemination.

DOI: 10.4018/978-1-5225-1871-6.ch002

INTRODUCTION

Medical informatics, health informatics, clinical informatics and bioinformatics are newly emerged disciplines in the subject tree, where application of Information Technology (IT) in management of information constitutes a major activity. There are various debates going on amongst the health, medical, information and computer professionals about the scope, activities and support to these fields through application of IT. The development of information systems to support the infrastructure of medical, understanding the needs of health professionals, managing data generated through clinical practices are some of the issues emerged which laid the foundation for supporting education, decision making, and communication. In this regard the emergence of Medical informatics recorded as a field which is concerns itself information professing for helping the tasks of medical practice, education, and research with the help of Information and communication Technology (ICT). Simultaneously, Health Informatics and Clinical Informatics are being used to cover the various aspects of informatics whereas bioinformatics is the convergence of biology for research, development, or application of computational tools and approaches for expanding the use of biological, medical, behavioral or health data, including those to acquire, store, organize, archive, analyze, or visualize such data (Huerta 2000). Historically, the medical informatics originated about 50 years back when systematic approaches applied for processing of data information and knowledge related to medicine and healthcare. A detailed discussion on the origin of the medical informatics was first reported by Collen in 1986 (Collen, 1986).

The biggest leap in the field of 'informatics' achieved with the emergence of 'Bioinformatics' as new discipline after completion of 'Human Genome Project' (HGP) in 2003 (Lim, 2000). The completion of HGP is supposed to be the main root for the origin of bioinformatics. The word conceptualized in 1980s, flourished in 1990s and become one of the major data extensive fields of research in last decade. The completion of HGP poses new challenges in front clinicians, scientists, doctors and researchers as huge amount of data generated in the form of sequences of organism and plants.

Both Medical Informatics and Bioinformatics now playing crucial role in the study related to genome, developing diagnostic test, updating genetic and medical data in clinical practice. At the same time libraries and information centers are shifting their roles and adopting the technological advancement fueled by information technology, bioinformatics, and networked information. These developments are changing both the role of library as well as context of information delivery systems for the subject (medical informatics, bioinformatics) as well as people (doctors, clinicians). The changes have reflected in the library various library activities and reflected the how patients, health care providers, researchers, policymakers, and the general public all relate to the corpus of biomedical information (Geer, 2006). Before moving further, it is essential to understand the domain of bioinformatics.

BIOINFORMATICS AND ITS RESEARCH DOMAIN

The bioinformatics research area involves the in-silico analysis of genomics data generated though laboratory experiments. Biological data are being produced at an unprecedented rate on a daily basis conjuring up in managing and integrating these data in a more precise and meaningful manner. It can be evident from the exponential growth in the database in every year. As an example the current version of European Molecular Biology Laboratory (EMBL) database consists of 199,575,971 entries with 301,588,430,608 nucleotides (Leinonen et al., 2010). After the completion of Human Genome Project in 2003 and the result obtained though this project has revealed and opens several doors of science and computation research in a variety of subject areas including bioinformatics. Completion of human genome project has generated immense interest to biotechnology industry and pharmaceutical companies. The research focus includes diagnosis of genes and their altered functions associated with diseases, identification of therapeutic targets and designing drugs against them, development of biomarkers for diagnosis and prognosis, etc. at individual level. Besides, genome sequences of many more organisms are available and are accessible publicly which has generated tremendous interest among the biologists for translational research throughout the globe.

As a result, the research outputs across the world are very much dispersed, redundant and available in different formats. Hence the vast amount of biological information is in a desperate need to be consolidated, stored, organized and indexed in a user friendly platform for better accessibility and usage. The existing available information in biological science is mind-boggling. It has been estimated that there are about 30,000 genes in the human genome that code for the 1.5 millions of protein. Each gene requires approximately 300 terabytes of trace file so that the total data to be handled in the tune of 30,000 times 300 terabytes. According to some estimates, medical imaging itself generates 400 million gigabytes of data annually (Karsch-Mizrachi & Ouellette 2001).

The bioinformatics is multidisciplinary in nature and applications of bioinformatics can be looked at the three level. At first level it is used to organize biological data to help the researchers' access information, add new information arising out of experiments and modify existing information of data sets such as genome sequences, macromolecular structures, and data from functional genomics. At second level, the researchers work on development of tool and resource that aid in the analysis of data. At third level bioinformatics finds its application in the use of tools to analyze the data and interpret the result in biologically meaningful manner.

All these three activities involve different activities of information search and retrieval every day and these activities are named as sequence comparison, linkage analysis, phylogenetic analysis, sequence assembly, data mining, sequence prediction, drug discovery and so on.

As result of bioinformatics research at this point of time the information in bioinformatics has become a crucial point of concern. At this juncture it is now essential to understand the information resources generated through various bioinformatics projects. The application of IT in biology for generating new hypothesis have yielded different product – better information product. The section ahead discusses the emergence of different bioinformatics information resources.

INFORMATION ACTIVITIES IN BIOINFORMATICS

The coverage of bioinformatics has been expanded into various subsidiary subjects. The expansions can be visualized as nucleic acid and genome sequence analysis, micro array analysis, protein sequence analysis, medical informatics, database building, drug discovery, etc. (Rastogi, Mendiratta, & Rastogi, 2006). Enhanced research activities in these disciplines have generated huge amount of data. This data is related to various sub topics/themes of Bioinformatics such as genome sequences, protein sequences, literatures, etc. and these are available over internet in different forms. Bioinformatics deals with computational analysis of these data. The scope of the Bioinformatics research activity includes the following five key activities:

1. **Data Acquisition:** The collection of data generated directly by laboratory experimentation.
2. **Database Development:** The collection of data of a single/group of topics/themes of bioinformatics in the computer-based repositories, which allows the efficient searching, retrieval and examination of the data.
3. **Data Analysis:** The processing of raw data into usable form.
4. **Data Assimilation:** The preparation of grid data for forecasting and modeling of data which results in generation of information.
5. **Analysis of Assimilated Data for Further Research:** The gathering, integrating and combining of the data with other topics. This process occurs repeatedly and creates a *"knowledge base"*. This knowledge base is basically the products such as databases, software tools, web servers, etc. (Wang, Zou and Zhu 2000).

INFORMATION RESOURCES IN BIOINFORMATICS

As the field of bioinformatics growing in its activities, scope and coverage, the researchers have a continuing and growing need for access to state-of-the-art bioinformatics and biological resources. The categorization of the resources depends upon its use and application for the community. On the basis of activities and products, the bioinformatics resources are categorized as the following:

1. Literatures,
2. Databases,
3. Web servers, and
4. Software tools.

Other than these are also resources which are available in the form of sequence data for both animal and plants and referred as raw data for experiments and when these raw data processed and becomes a resource for future use.

Bioinformatics Literatures

The increased research activity has witnessed generation of huge data and information. This information is utilized in various research related activities/applications. This information is brought out as information products / publications. An online search was conducted in NCBI-PubMed to elucidate the growth of literature using '*Bioinformatics/Computational Biology*' (both the terms are synonymous). It is found that there are 98882 publications. In 1990, there were only 126 publications and it reached to 98882 by the 2011. It can infer that the cumulative growth is approximately 800 times within a period of 21 years (on an average growth of approximately 38 times per year) (PubMed).

Bioinformatics Databases

Databases are the one of the prominent resources in the area of bioinformatics. All these activities are undertaken in one phrase 'Biological Database Management System' leading to a product known as 'Databases' rather 'Bioinformatics Databases'. The completion of HGP revolutionized the database activities. With the availability of new technologies researchers start using them for different activities which resulted in rapid growth of the publications. Several journals started accepting papers with research data that stored in databases (centralized repositories). These databases served as a warehouse of the information that has been published in the journals/ reports. These databases are easily available and accessible due to advancement of

ICT. There was a need felt to establish database management system which deals the effective management of information. The National Center for Biological Information (NCBI), GeneBank, European Molecular Biology Laboratory (EMBL), and DNA Data Bank of Japan (DDBJ) are storing, organizing and maintaining these databases.

The efforts are being made to consolidate and conglomerate the databases and in this line Nucleic Acid Research Journal published by Oxford University Press bringing a special issue every year dedicated to the databases (Fernandez-Suarez & Galperin, 2012). The classifications of these databases are available in fifteen different categories encompassing the different area of bioinformatics, as shown in Table 1. There is a multifold growth in the research output from last two decades. During this period new databases has been created and identified resulting in formation of new categories of databases. In the year 2011 a new category namely 'Cell Biology' has been created. The category cell biology is nascent category with only four databases. Whereas 22 years old genomic database category has maximum number databases *i.e.* 339 databases. It can be also seen that 19 years old category

Table 1. Number of databases developed in different areas of bioinformatics

Categories of Databases	1989-1994	1995-1999	2000-2004	2005-2009	2010-2011	Grand Total
Cell Biology	0	0	0	0	4	4
Genomics Databases (non-vertebrate)	2	10	61	178	88	339
Human and other Vertebrate Genomes	1	4	32	57	35	129
Human Genes and Diseases	1	17	46	92	83	239
Immunological Databases	0	1	5	19	6	31
Metabolic and Signaling Pathways	1	3	19	83	81	187
Microarray Data and other Gene Expression Databases	0	2	18	24	20	64
Nucleotide Sequence Databases	1	9	36	72	20	138
Organelle Databases	0	1	7	15	5	28
Other Molecular Biology Databases	0	2	4	43	23	72
Plant Databases	0	4	35	70	29	138
Protein Sequence Databases	2	14	69	111	46	242
Proteomics Resources	0	0	2	12	3	17
RNA sequence databases	0	3	21	44	19	87
Structure Databases	5	5	34	65	43	152
Grand Total	13	75	389	885	505	1,867

protein sequence database has second highest (242databases) and the category Human genes and disease, which old by 18 years, ranked third with 239 databases. It can be concluded that the longer is the period of research in particular category more is the number of databases.

The first ten years during 1989 to 1998 the growth of databases has increased from only one to 56, *i.e.* an annual growth of about 6 databases. Whereas during the decade 1999 to 2008 the number of databases has grown from 56 to 1225 databases, which is annual growth of about 117 in comparison to earlier decades. During the last three years *i.e.* 2009 and 2011 there is an addition of another 642 databases, making it 1867 databases with an annual growth of 160 databases per year. It can construe that the research, research output and literature has multiple growth during last few years. It also indicates increased research year after year.

Bioinformatics Web Servers

The researchers search for updated and relevant data/information using one or more search engines. The search engines may or may not provide links to the relevant data/information. Linking to irrelevant sites crate more confusion and complex situation and research may not progress in right direction. To overcome this unwarranted situation, the cutting-edge 'data analysis software packages' have been developed. These packages search the most up-to-date versions of databases. These data analysis software packages analyses the myriad datasets spread over different database system through chains of computers, commonly known as web servers.

The web browsers through Internet initiate a connection to the Web servers, which store the web pages and convert the domain name into an IP address. The Web server stores all the files that display contents of the websites. The web servers maintain and provide links to useful content available over Internet. Every specialized web server provides the facility for using the software/tools for different purposes such as analysis, modeling, comparison and prediction of research data. Each web server may have one or more software programs. These programs may or may not be available on other web servers. This software can be either general or specialized. However, each web server performs number of activities through the available software and each one of them is categorized based on the major activity in specialized field of study/research.

These web servers have been broadly categorized into eleven groups based on their activities. The categorization is by the subject (Protein), field of activity (Education), and mode of dissemination (Literature). These broad categorizations have been made by Nucleic Acids Research Journal which publishes the details of web servers in one of its issues published every year in the month of July (Benson,

2013). Table 2 presents a scenario of different categories of web servers available globally for the purpose of computational analysis.

It can be observed from Table 2 that the three major fields i.e. Protein (1214), DNA (585) and Expression relation (427) web servers are ranked (in terms of numbers) as number one, two and three respectively. This number of web servers is also increasing year after year.

Software Tools

The database contains massive amounts of experimental data. Browsing and analyzing the data is fascinating and will surely lead to many interesting discoveries. Searching for relevant information in bioinformatics databases is a critical part for the development of research projects. A search through bioinformatics databases often initiated by posing a query related to biological problems. To exploit and experiment this data, computational and algorithmic methods are designed. These algorithms ultimately lead to the use of computational software for the analysis of data generated through biological and biochemical experimental processes. These methods are used for testing the data and creation of new theories. Using the ICT, software is designed and developed for conducting the process of collecting, storing and analyzing the growing datasets, which results in creation of new database and discovery /establishment of the new arena for the subject. Some of the example of

Table 2. Number of web servers developed in different areas of bioinformatics

Sr. No.	Subject Category	Total Number of Web Servers					
		2006	2007	2008	2009	2010	2011
1	Computer Related	70	64	64	67	76	82
2	DNA Related	279	329	345	366	512	585
3	Education Related	72	76	75	75	73	73
4	Expression (Genes, Proteins)	224	272	309	350	393	427
5	Human Genome	107	129	137	151	177	214
6	Literatures	29	35	42	51	54	60
7	Model Organisms	181	202	215	229	258	361
8	Other Molecules	9	15	18	28	44	76
9	Protein Related	654	795	850	954	1044	1081
10	RNA Related	92	107	123	134	144	184
11	Sequence Comparison	171	244	256	272	299	307
Total							3450

the software tools are BioPerl, BioJava, and EMBOSS. The use of the software is different in case of different activities involved in bioinformatics industry. Bothe open as well as proprietary software tools are now available for different purposes.

A number of other resources are increasing which includes patents, dedicated organizations, monographs, and people. The need is to identify each resource, its people and treat them as a resource.

PROBLEMS OF RESOURCE INTEGRATION

Based on the analysis it is found that the Bioinformatics mostly deals with the database and tools as a methodology for the management of biological data, information, and knowledge. The rapid development of the subject and massive generation of the information in bioinformatics poses different problems associated with identification, selection, consolidation and integration for a cohesive access. The access information issue is related to the following:

1. Understanding the range of bioinformatics user and their information needs,
2. An understanding of how these users seek information and use resources,
3. A variety of resources and their file format than traditional information resources, and
4. The role of institutions in providing tailor made information services to the target user issues (Lynch, 1999).

METHODOLOGY FOR SYNERGY BETWEEN INFORMATION AND DISCIPLINE

Different person analyzed these problems and proposes the solution for access of information in bioinformatics. Macmullen and Denn (2005) analyzed the Information problem in bioinformatics while Bartlett and Neugbauer (2008) discusse the task based information retrieval for bioinformatics activities. Antezana *et al.* (2009) proposed a RDF integrated repository 'BioGateway' for integration of resources, whereas Guillermo de la Calle *et al.* (2009) proposed 'BIRI" and automatically discovering and indexing available public bioinformatics resources from the literature. Each system has some issue related to either crude bioinformatics activities or lacking at some point where pure bioinformatics resources intermingled and leads to the redundancy of results.

A Bioinformatics Reference Model proposed for developing framework for organizing bioinformatics resources (Hiew & Bellgard, 2007). Ram *et al.* (2010)

proposed clustering techniques for the organizing resources. Access to bioinformatics resources, available in different forms and format, are currently unevenly distributed over the internet and some are at private domain. The published literature provides details of the information and these are also scattered. The description of the resources does not meet the standard classification system to describe the resources, leading to ambiguity in search result. Sometimes the published literature does not contain reliable references to the location and to the availability of most bioinformatics resources (Cannata, Merelli & Altman, 2005). The only solution people find is use search engine like Google, whose results are based on page ranking. Because of lack of standardization, it is difficult to locate specific bioinformatics resources and finding exact information becomes very difficult.

A gap is created in between the technical works of bioinformatics (tool development, software development, data analysis) and resources needs of the user (journals, patents, databases) due to various factors, as seen in Figure 1. These factors may be information literacy, heterogeneity, technological barrier, format, etc. The technology in technological view is not in a form that can be naturally linked to the process of scientific views. This is because of different reasons associated with the format, size, and availability, functions, operations, and component and information literacy.

Figure 1. The gap created between the technical works of bioinformatics and the resources needs of the user

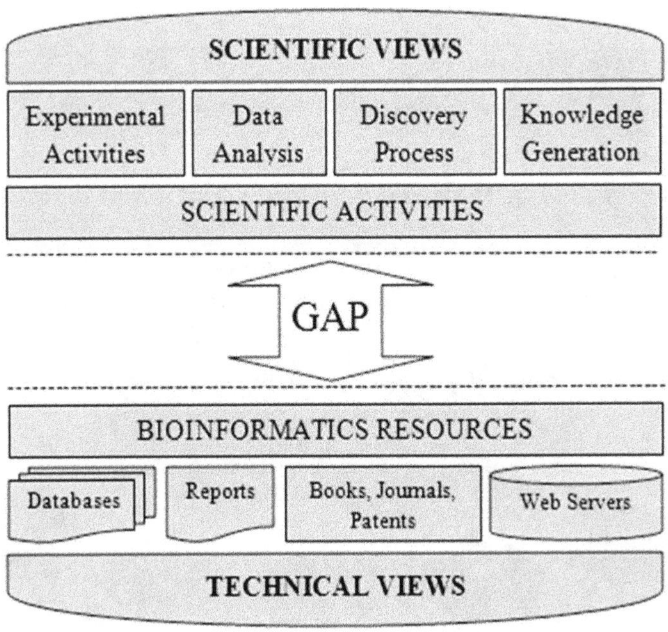

Under this situation if this gap is left unfilled, the processed bioinformatics resources may either be too technical and lack scientific relevance, or they may be built on scientific relevance but may lack technical qualities to use and operate effectively. The gap, which has been created between the two views above, fails to meet the objectives of searchability, integration and access. This gap can be fulfilled by providing an effective mechanism.

These mechanisms can be information literacy, creation of interactive web interface, or integration of technology into research relevant component and resources. In this way the delivery of information can be ensured for relevant component into the appropriate form for usage in the research process shown in Figure 2.

To overcome this problem a model has been developed keeping the view of the integration of various information resources in order to full fill the gap arises between scientific and technical views. The integration framework of model is given in Figure 2. The schematic framework conceptualized as *i*BIRA (Integrated Bioinformatics Information Resource Access) and implemented in the following manner (Ram & Rao 2012):

1. **Identification, Selection, and Classification of Information Resources:**
    a. The resources identified in the above discussed categories are based on their characteristics into six major categories:

Figure 2. Integrated bioinformatics information resource access framework for iBIRA

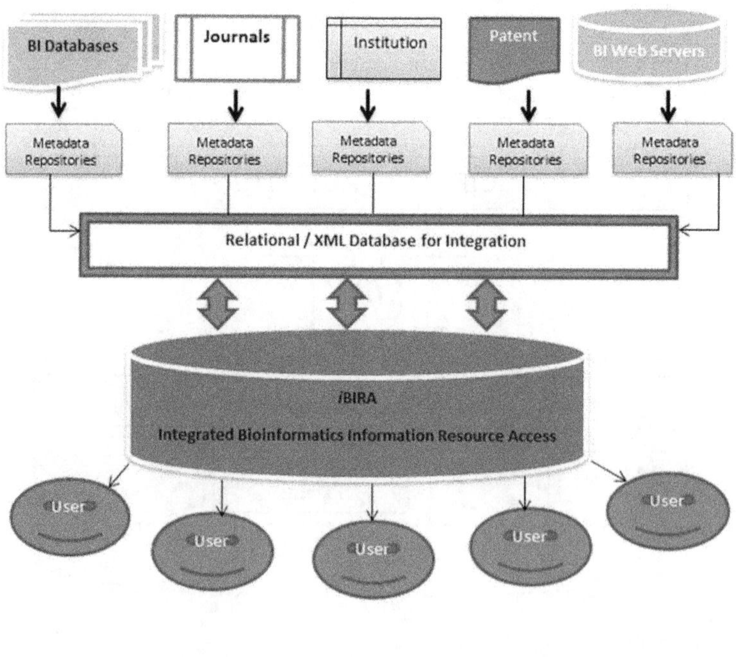

i. Databases,
ii. Web servers,
iii. Software tools,
iv. Journals,
v. Patents, and
vi. Institutions.

b. Further sub-categorization was made based on the utilities, activities and tasks. Major database source such as PubMed, Scopus, Web of Science, Journals individual web pages and Institutional websites were searched for the pertinent information. A resource hub was created based on the identified resources. For the description and standardization of the resources, the fifteen basic elements of the Dublin Core Metadata Initiative have been adopted and where ever necessary it is modified to suite the description (Dublin).

2. **Design and Development of Model:**

a. The design and development of the tool has been done based on software engineering approaches. Internally it is a three layered architecture.
i. Presentation Layer (the interface),
ii. Business layer (the search Schema), and
iii. Data access layer (the database of the model).

b. The interface has been designed with the help of PHP; a search algorithm is developed taking care of all the permutation and combination of user's perception as well as Boolean operators. The database is designed with the help of SQL.

c. The designed and developed model in this way obtained was tested at local server and after successful testing and verification, it is hosted over the internet for global use. See Figure 3.

DISCUSSION

For many years, medical informatics and health informatics are dealing with the wide range of applications and developing various clinical applications for supporting health community. These developments help community with better management of information. The application of Medical Informatics is now having gain momentum and being practiced in various countries (Maojo *et al.,* 2001; Collen, 1995). Similar to medical informatics, Bioinformatics is now gaining momentum in serving with clinical information through its activities, tools and techniques. The information problem associated with the integration of the information has become a challenge for the scientific community.

Figure 3. Integrated bioinformatics information resource access (iBIRA)

The design and development of the tool like *i*BIRA has a potential benefits and helping scientific community to find pertinent information related to bioinformatics and clinical practices in the form of databases, web servers, software tools, etc. The tool has integrated around more than five thousand resources in different area of the bioinformatics. Simultaneously, the newer and newer data are being updated on daily basis in order to make it more robust system of information hub.

The research can search bioinformatics information in six different categories, the bibliographic description of the resource, any pertinent literature available on the resource, key person responsible for the development resource and organization participated in the development. It also provides the link to the original web page from where more details can be found. The search criteria have multilevel queries where the user can combine it with different search criteria. The result obtained in this way is represented in the form of a tree structure as seen in Figure 4, where a user can easily understand the level of description.

Figure 4. Tree view obtained through a search page of iBIRA for specific queries

CONCLUSION

The tool iBIRA integrates different bioinformatics information resources and the data has been described as per the DCMI terms. The information resources are displayed in a helpful sequence in the form of tree view. There are several initiatives taken in order to consolidate the information, but those applications somewhere and at some point lacks integrated framework. The description of resources available through this tool could be beneficial to the wide range of community in identification, consolidation and providing access at single platform. There is a possibility as well as opportunity for further expansion of the tool with addition of newer and newer kinds of information resource and expanding its scope to cover wide range of applications.

REFERENCES

Antezana, E., Blondé, W., Egaña, M., Rutherford, A., Stevens, R., De Baets, B., & Kuiper, M et al.. (2009). BioGateway: A semantic systems biology tool for the life sciences. *BMC Bioinf.*, *10*(Suppl 10), S11. doi:10.1186/1471-2105-10-S10-S11 PMID:19796395

Bartlett, J. C., & Neugebauer, T. (2008), A task-based information retrieval interface to support bioinformatics analysis. *Proceedings of International Symposium on Information Interaction in Context*, (pp. 97-101).

Benson, G. (2013). Nucleic Acids Research Annual Web Server Issue in 2013 – Editorial. *Nucleic Acids Research*, *41*(W1), W1–W2. doi:10.1093/nar/gkt559

Cannata, N., Merelli, E. & Altman, R.B (2005), Time to organize the bioinformatics resourceome. *PLoS Computational Biology*, *1*, 531-533.

Collen, M. (1995). *A history of medical informatics in the Unites States: 1950 to 1990*. Bethesda, MD: American Medical Informatics Association.

Collen, M. F. (1986). Origin of Medical Informatics. *The Western Journal of Medicine*, *145*, 78–86. PMID:3544507

de la Calle, G., García-Remesal, M., Chiesa, S., de la Iglesia, D., & Maojo, V. (2009). BIRI: A new approach for automatically discovering and indexing available public bioinformatics resources from the literature. *BMC Bioinf.*, *10*(1), 320. doi:10.1186/1471-2105-10-320 PMID:19811635

Dublin Core Metadata Initiatives. (n.d.). Retrieved from http://dublincore.org/documents/dcmi-terms/

Fernandez-Suarez, X. M., & Galperin, M. Y. (2012). The 2013 Nucleic Acids Research Database Issue and the online Molecular Biology Database Collection. *Nucleic Acids Research*, *41*(D1), D1–D7. doi:10.1093/nar/gks1297 PMID:23203983

Geer, R. C. (2006). Broad issues to consider for library involvement in bioinformatics. *Journal of the Medical Library Association: JMLA*, *94*(3), 286–303. PMID:16888662

Hiew, H. L., & Bellgard, M. A. (2007). Bioinformatics Reference Model: Towards a Framework for Developing and Organising Bioinformatic Resources. In *International Symposium on Computational Models of Life Sciences*. doi:10.1063/1.2816640

Huerta, M. (2000). *NIH working definition of bioinformatics and Computational biology*. Retrieved from http://www.bisti.nih.gov/docs/CompuBioDef.pdf

Karsch-Mizrachi, I., & Ouellette, B. F. F. (2001). The Genbank Sequence Database. In *Bioinformatics: A Practical Guide to the Analysis of Genes and Proteins* (pp. 45–63). John Wiley & Sons, Inc.

Leinonen, R., Akhtar, R., Birney, E., Bower, L., Cerdeno-Tarraga, A., Cheng, Y., & Cochrane, G. et al. (2010). European Nucleotide Archive. *Nucleic Acids Research, 39*(suppl1), D28–D31. doi:10.1093/nar/gkq967 PMID:20972220

Lim, H., & Venkatesh, T. V. (2000). Bioinformatics in the pre- and post-genomic eras. *Trends in Biotechnology, 18*(4), 133–135. doi:10.1016/S0167-7799(99)01409-2 PMID:10809530

Lynch, C. (1999). Medical libraries, bioinformatics, and networked information: A coming convergence? *Bulletin of the Medical Library Association, 87*, 408–414. PMID:10550026

MacMullen, W. J., & Denn, S. O. (2005). Information problems in molecular biology and bioinformatics. *Journal of the American Society for Information Science and Technology, 56*(5), 447–456. doi:10.1002/asi.20134

Maojo, V., Iakovidis, I., Martin-Sanchez, F., Crespo, J., & Kulikowski, C. (2001). Medical Informatics and Bioinformatics: European Efforts to Facilitate Synergy. *Journal of Biomedical Informatics, 34*(6), 423–427. doi:10.1006/jbin.2002.1042 PMID:12198762

Ram, S., & Rao, N. L. (2012). *i*BIRA – integrated bioinformatics information resource access: Organizing the bioinformatics resourceome. *Reference Services Review, 40*(2), 326–343. doi:10.1108/00907321211228354

Ram, S., Sureka, R. K., Sharma, P., & Rao, N. L. (2010). Integrating bioinformatics information resources: Management of information through clustering. In *IEEE Students'Technology Symposium* (TechSym).

Rastogi, S. C., Mendiratta, N., & Rastogi, P. (2006). *Bioinformatics: Methods and Applications – Genomics, Proteomics and Drug Discovery*. New Delhi: Prentice Hall of India.

Wang, B., Zou, X., & Zhu, J. (2000). Data Assimilation and its Applications. *Proceedings of the National Academy of Sciences of the United States of America, 97*(21), 11143–11144. doi:10.1073/pnas.97.21.11143 PMID:11027322

Chapter 3
Information Needs of Bioinformatics Researchers

Manlunching
Saha Institute of Nuclear Physics, India

ABSTRACT

Information plays a vital role in bioinformatics to achieve the existing bioinformatics information technologies and to identify the needs of bioinformatics researchers. The most revolutionary development for bioinformatics resources is access to the internet because internet is pervasive in all bioinformatics work. Users required various sources of information for conducting bioinformatics research. The success of the information service is more likely to be achieved by adjusting the services to meet the specific needs of an individual.

INTRODUCTION

Library and information science focus on information seeking and the information user, while those from the field of communications focus on the communicator and the communication process (Robson & Robinson, 2013). Needs may refer to lack of self-sufficiency and also represents gap in the present knowledge of the users. Apart from the expressed or articulated needs, there are unexpressed needs which the user is aware of but does not like to express consciously or unconsciously (Devadason & Lingam, 1996). Information is used, in the context of user-studies

DOI: 10.4018/978-1-5225-1871-6.ch003

research. There is not much effort in research and writing of user studies that has circumstances in information science apart from information retrieval. The probable interrelationships among personal needs and other factors aim is to suggest that when we talk about users' information needs we should have in mind some conception of information (facts, data, opinion, advice) as one means towards the end of satisfying such fundamental needs. Information needs should not be confused with information seeking behavior. What users believe they need is represented in the subjective understanding of needs. This subjective understanding is reflected in their information seeking behavior. Even if this behavior may be studied objectively it is still not useful as criteria for what is needed. What is needed is something that is able to solve the problem behind the users' behavior (Wilson, 1981). Information plays a vital role in bioinformatics to achieve the existing bioinformatics information technologies. Information is recognized as a national resource, which is of vital significance in all sectors of human endeavor - planning, decision making, research and development, education, socio-economic and cultural development, and also in improving the quality of life of every members of the society. Along with the material and energy, information is considered a potential resource, a product and there by a need, which must be put to use effectively. It is true that the information scientists had for a long time neglected one of the most important components of any information system, namely the 'user'. They were more concerned with the information and their bibliographical organization and control. How exactly the user behaved when he was looking for some information, what type of information was used in which situation, how the information was used when obtained, all these were not very clearly known to the information scientists. Proper systematic planning and development of information resources and services of the user studies are very essential. In recent years, there have been several studies pertaining to bioinformatics researchers and their information needs in bioinformatics resources. However more need in bioinformatics resources has come to pass and the author discussed some topics in this chapter to get the unambiguous inspiration.

DEFINITION OF BIOINFORMATICS

Paulien Hogeweg coined the term bioinformatics in 1970 for the study of informatics processes in biotic systems. 'Bio' means Molecular Biology and 'Informatics' means Computer Science. The study of the application of molecular biology, computer science, artificial intelligence, statistics and mathematics, organizes, understand and discover interesting information associated with the large-scale molecular biology databases and to guide assays for biological experiments is known as Bioinformatics (Gilbert, 2007). Bioinformatics is the field of science in which biology, computer science, and information technology merge to form a single discipline. The ultimate goal of the field is to enable the discovery of new biological insights as well as to create a global perspective from which unifying principles in biology can be discerned. In another words, Bioinformatics is the design and development of computer-based technology that supports life science. Using this definition, bioinformatics tools and systems perform a diverse range of functions including: data collection, data mining, data analysis, data management, data integration, simulation, statistics, and visualization (Lacroix & Critchlow, 2003).

HYPOTHESES

1. The information needs and uses in bioinformatics will only increase if the level of new bioinformatics systems grows.
2. Information services assist users in identifying and utilizing of bioinformatics tools.
3. Staff development programme for bioinformatics users enable them to develop multidisciplinary skills.
4. Bioinformatics services of the library/ centre will greatly depends on the level of available subject specific expertise.

INTERNET ACCESS

Users begin their research on the internet much like any other information seeker, consult their faculty advisors before other people, and use libraries in diverse ways depending on the discipline they studied (Catalano, 2013). The most revolutionary development for bioinformatics resources is access to the internet because internet is pervasive in all bioinformatics work. The web browser has become the primary means of access to bioinformatics resources for most bioinformatics researchers. Primary databases as well as derived databases created by the analysis and/or annotation of this information and software tools are available over the web. It is not practical to maintain local copies of complete, up-to-date sequence databases and tools on a personal computer (Brown, 2000). Researchers access internet for different purpose based on their requirements of what kind of information they urge for. The purposes of accessing internet are: E-Resources, Keeping self-update, Reference and others depending on their research purpose. Those users who use internet access often are more self-update than those who rarely use for accessing bioinformatics resources. This is true in case of accessing online books and journals in which accessing those online resources needs internet access since all the online resources are available in internet through their specified registered Internet Protocol address. In today's world Internet is widely used by researchers in and outside the bioinformatics centers to meet the requirements of their research (Manlunching, 2014).

FACILITIES REQUIRED

In this digitized world and India being a developing country, most of the libraries are on their way to digitization and subscription of electronic resources is the current trend in libraries which is also cost effective. This results to saving space, saving time of the users and library personnel's'. In addition to the existing facilities bioinformatics researchers are asked what more additional facilities are required by them to meet their needs and requirements and to improve the library facilities. Strong interest was expressed in additional facilities required as shown below:

1. Increase holdings of laboratory manuals, handbooks, and other reference material.
2. Provide more online access to other databases and reference sources.
3. Provide networked access to current contents.
4. Establish a collection of research laboratory templates.
5. Provide access to more full-text online journals and books.

Here the argument is that libraries must maintain a well-rounded core collection development, including reference material to satisfy the information needs of the bioinformatics researchers. These may be supplemented through networks, e-resources, library consortium, etc., to achieve better qualitative and quantitative standards. Library collections are dynamic resources therefore, constant renewal of library materials and library collections to ensure that the collections remains relevant to the users is essential (Manlunching, 2014).

SOURCE OF ACQUIRING BIOINFORMATICS RESOURCES

Users required various sources of information for conducting bioinformatics research. Study conducted by Manlunching (2014) finds that the source of acquiring bioinformatics resources are categorized in nine parameters as enunciated below:

1. Reading of review articles.
2. Workshop, conference, symposium, etc.
3. Own research.

4. Access to sequence analysis software.
5. Accessing of bibliographic databases from library.
6. Printed and electronic media.
7. Personal contacts with other researchers or discussion with colleagues.
8. Access to library services.
9. Contact and discussed with information professionals in geographic areas through video conference and or online chatting.

Researchers required immense sources of information inside and outside of their study zone. Yet this is no easy task, and would require more resources for processing and improvement of context based knowledge sources, to develop new research. In fact, through social networking technology, they can ask the global community to answer their queries (not just librarians, but the subject experts in each field who are ready to answer) especially when the question is specific and the expected answer is not more than a few sentence.

OPINION ABOUT BIOINFORMATICS SERVICES AND RESOURCES

Rating refers to expression of judgment or opinion grading a phenomenon (an object or person) in terms of specified criteria. The users are expected to rate each elements (Agree, Depends and Disagree) separately in terms of specified criteria by selecting a numerical rating that is offered to them which are presented in Table 1. These opinions are the main hypotheses framed for the study of the information needs of bioinformatics researcher (Manlunching, 2014).

The Chi-Square Test in Figure 1 is used to determine whether there is a significant difference between the expected frequencies and the observed frequencies in one or more data and to determine the accurate responds.

Now, applying the Chi-Square formula for finding the Chi-Square value (X^2) and find out if there is a significant difference between the observed and expected opinion. After calculating the X^2, we are now finding the *df* (degree of freedom). *df* refers to the number of values that are free to vary after restriction has been placed on the data and is defined as 'n-1'. Here, our *df* = n-1; i.e. 3-1= 2; and our P value (Critical value) = 0.05. Now, using the table value for Chi- Square let's find the *df* and critical value of P *(Chi-Square table value attached in Appendix)*. The table value for Chi-Square in the correct box of *2 df* and *P = 0.05* level of significance is 5.99. If the Chi-Square value (X^2) is greater than the table value (5.99 i.e. $X^2 > 5.99$) after calculating, that means the hypothesis must be rejected, but if the X^2 value is smaller than the table value (5.99 i.e. $X^2 < 5.99$) the hypothesis should be

Table 1. Opinions about bioinformatics services and resources

Hypotheses	Criteria	O	E	O - E	(O - E)^2	(O - E)^2/E
1. The information needs and uses in bioinformatics will only increase if the level of new bioinformatics systems grows	Agree	300	290	10	100	0.34
	Depends	10	15	-5	25	1.67
	Disagree	5	10	-5	25	2.50
	X^2 = 4.51					
2. Information services assist users in identifying and utilizing of bioinformatics tools	Agree	297	300	-3	9	0.03
	Depends	15	10	5	25	2.50
	Disagree	3	5	-2	4	0.80
	X^2 = 3.33					
3. Staff development programme for bioinformatics users enable them to develop multidisciplinary skills	Agree	175	180	-5	25	0.14
	Depends	121	120	1	1	0.01
	Disagree	19	15	4	16	1.07
	X^2 = 1.22					
4. Bioinformatics services of the library/ centre will greatly depends on the level of available subject specific expertise	Agree	180	200	-20	400	2.00
	Depends	107	90	17	289	3.21
	Disagree	27	25	2	4	0.16
	X^2 = 5.37					

Figure 1. Chi-square value formula
O is the Observed Frequency in each category (i.e. total no. of respondents).
E is the Expected Frequency in the corresponding category is sum of (i.e. expected response from the user).
df is the "degree of freedom" (n-1).
X^2 is Chi Square Value.

$$X^2 = \frac{(O - E)^2}{E}$$

accepted. When the first hypothesis is framed, the expected response from the users is 290 (which is decode as 'E' meaning expected frequency) but after distributing the questionnaire to the users, the response received is 300 ('O' means observed frequency/response received from the respondents). Similarly, the expected and observed frequency for the second, third and fourth hypothesis are shown in Table 1. The study reveals that:

1. The first hypothesis testing Chi-Square value i.e. $X^2 = 4.51$ is less than the level of significance i.e. 5.99 (4.51< 5.99). Hence, the hypothesis 'the information needs and uses in bioinformatics will only increase if the level of new bioinformatics systems grows' is accepted. The users understand and identify the services and availability of the existing bioinformatics system grows and they have declared their additional services required from the library such as: increase holdings of laboratory manuals, handbooks, and other reference material, provide more online access to other databases and reference sources, provide networked access to current contents, establish a collection of research laboratory templates, and provide access to more full-text online journals and books. In addition to the additional services, the bioinformatics users state the frequency and purpose of performing internet access for availing the bioinformatics services and resources. Due to the overload of information, bioinformatician faced major challenges when tracking down new discoveries and the results of research in their domain of interest. These challenges are intensified by the need to follow developments in other domains that might possibly be relevant to one's own research.
2. Similarly, the second hypothesis 'Information services assist users in identifying and utilizing of bioinformatics tools' is accepted as the Chi-Square value ($X^2 = 3.33$) is less than the level of significance i.e. 5.99 (3.33 < 5.99). Performing internet access for E-Resources, Keeping self-update, Reference and depending on their research purpose and the types of information used such as; Review full text articles on particular subject, State-of-the-art review article, Authentic source of information and using multimedia files utilized in identifying bioinformatics tools.
3. Likewise, the third hypothesis 'Staff development programme for bioinformatics users enable them to develop multidisciplinary skills' is accepted due to the Chi-Square value ($X^2 = 1.22$) is less than the level of significance i.e. 5.99 (1.22 < 5.99). The workshops, seminars, conferences and continuing education and internet clearly reflected intense interest in the database connections, and the need for structures to be exploited further to facilitate use of the massive amount of biological data being encoded. Users access bioinformatics services for different purpose based on their requirement of what kind of information they urge for. They used internet for e-resources, keeping self-update and reference. Due to these technological advances, users can try to find out themselves the information they need. But they must spend some time on their networked computers, patiently scan through the information retrieved, and sift/filter to get the most relevant data. Users' source of acquiring bioinformatics resources are; Reading of review articles, workshop, conference, symposium, etc., own research, access to sequence analysis software, accessing of bibliographic

databases from library or laboratory computers, printed and electronic media, personal contacts with other researchers or discussion with colleagues, access to library services and contact with Information professionals in geographic areas.

4. Finally, the fourth hypothesis 'Bioinformatics services of the library/ centre will greatly depends on the level of available subject specific expertise' is partially accepted because the Chi-Square value $X^2 = 5.37$ and the level of significance i.e. 5.99 ($5.37 < 5.99$) is very meager in difference. Humphrey (1967) defines a subject librarian as "a member of a library staff appointed to develop one or more aspects of a library's technical or reference service in a particular subject field. Although he would already have some experience in his field and would commonly have obtained a first or a research degree in the subject, it is not essential that he should have qualifications in the subject when he is appointed". Many institutions, particularly in the United Kingdom have different designations or titles for subject librarians. They are known as faculty librarians, school librarians, subject consultants, subject support officers, subject specialists, academic librarians, liaison librarians, link librarians, information librarians, information specialist, or subject librarians. According to Martin (1996), the majority of librarians in Britain prefer to be known as either a "liaison librarian" or a "subject librarian". They believe that a subject specialist denotes that one has serious subject knowledge and qualifications, whereas a subject librarian is a professional librarian who happens to look after a particular subject. There is also a lack of requisite in-service training programme for librarians after their professional training. Continuous professional development should be career long and should match changes in personal circumstances with changes in organizational structures and job requirements. For subject librarians to be multi-skilled information personnel they require a variety of training needs which include: training in the management of change in order to respond to new roles, responsibilities and, in some cases, new locations; communication skills in order to work effectively with individuals, groups and committees; training in time management in order to plan and prioritize their increased workload; training in teaching and learning methods and skills to improve user education and training in the use and evaluation of electronic resources, especially the Internet.

From the testing of the hypotheses using the Chi-Square test, we can conclude that as the primary role of the library has shifted from information repository to portal, the need for specialized subject expertise has both broadened and intensified. Few librarians have the training or experience that would enable them to provide a full range of bioinformatics services.

SELECTION OF BIOINFORMATICS JOURNALS DATABASES IN LIBRARIES

The major failure of current bioinformatics systems is the inability to access and interlink all relevant journals databases. However librarian took initiatives in interlinking various databases by providing a consistent query method and hyperlink between related databases (Brown, 2000). In fact library is a non-profit organization with limited source of funds. Therefore librarians' attempt to select the bioinformatics journals database on user's point of view and taking their opinion based on the satisfaction of their needs and requirements. Study conducted by Manlunching (2014) depicts that "Cancel duplicate print subscriptions if electronic databases is available" and "Subscribe to only the electronic versions of new database titles" is agreed by 95% researchers. 94% researchers agreed to "Cancel lesser used print databases"; 80% agreed to "Place fewer new subscriptions to print databases"; 64% agreed to "Financed additional electronic databases by 'pay per use' (users pay per article accessed)" whereas 82% disagree to "Reduce the number of print books purchased in case of e-books availability". The analysis reveals that, electronic and digitized documents are more preferred by the users than print, although for reading purpose users choose the print form of document. Electronic journals databases are easy to access for retrieving full text article within a short period. It saves the time of the users in locating their desired information.

USERS TRAINING AND WORKSHOP

Knowledge gaps and extensive learning time engaged as key factors that in combination inhibited the use of bioinformatics tools (Shachak, Shuval & Fine, 2007). In particular, trainee felt that learning to analyze the results would require a substantial learning effort and time investment. Those users attended the training and workshops are more informative and update with the latest trends in bioinformatics than those who do not attend the training and workshop. The major areas of training and workshop attended by the users are; bioinformatics software and tools, computer aided instruction, modeling and simulation, handling and management of biological data, including its organization, control, linkages, analysis and so forth, search and retrieval of biological information, Bioinformatics and its emerging dimensions, Computational Biology, Genome analysis, protein structure prediction and drug design, and Routine sequence analysis (Manlunching, 2014). The maximum used of the bioinformatics resources and satisfying a query by researchers has been seems to be a question of how good the awareness programme was. Therefore alternative ways to analyze is to examine that training and workshop plays an important role for bioinformatics users.

CONCLUSION

Information is a keyword in today and tomorrow's society. A well-developed library system with relevant subject expert personnel is a condition for meeting the challenges of the information needs of bioinformatics researchers. In the present age of information, it has been increasingly felt that to serve users better, identification of the information needs must become the central focus of attention. It is beyond doubt that the success of the information service is more likely to be achieved by adjusting the services to meet the specific needs of an individuals or a specific group rather than trying to adopt the users to match with the output of the information system. Librarians have to create and develop the user-oriented system for their maximum information satisfaction.

Libraries must maintain a well-rounded core collection development, including reference material to satisfy the information needs and uses of the bioinformatics users. These may be supplemented through networks, e-resources, library consortium, etc., to achieve better qualitative and quantitative standards. Library collections are dynamic resources therefore, constant renewal of materials/ collections to ensure that the collection remains relevant to the users is essential.

REFERENCES

Brown, S. M. (2000). *Bioinformatics: a biologist's guide to computing and the internet*. New York, NY: Eaton Publishing.

Catalano, A. (2013). Patterns of graduate students information seeking behavior: A meta-synthesis of the literature. *The Journal of Documentation*, *69*(2), 243–274. doi:10.1108/00220411311300066

Devadason, F. J., & Lingam, P. P. (1996). A Methodology for the Identification of Information Needs of Users. In *62nd IFLA General Conference - Conference Proceedings*. Beijing, China: International Federation of Library Associations and Institutions. Retrieved August 22, 2009, from http://www.ifla.org/IV/ifla62/62-devf.htm

Gilbert, D. (2007). Bioinformatics Introduction. Bioinformatics Research Centre. Retrieved from www.brc.dcs.gla.ac.uk/~drg/.../bioinformaticsHM0607/slides/intro.pdf

Humphreys, K. (1967). The subject specialist and the national and University libraries. *Libri*, *17*(1), 29–41.

Lacroix, Z., & Critchlow, T. (Eds.). (2003). *Bioinformatics: Managing scientific data*. San Francisco: Morgan Kaufmann.

Manlunching. (2014). *Information Needs and the Uses of Bioinformatics Users of Select Libraries in India: A Study* (Unpublished doctoral thesis). University of Delhi, New Delhi, India.

Martin, J. V. (1996). Subject specialization in British University libraries: A second survey. *Journal of Librarianship and Information Science*, *28*(3), 159–169. doi:10.1177/096100069602800305

Robson, A., & Robinson, L. (2013). Building on models of information behaviour: Linking information seeking and communication. *The Journal of Documentation*, *69*(2), 169–193. doi:10.1108/00220411311300039

Shachak, A., Shuval, K., & Fine, S. (2007). Barriers and enablers to the acceptance of bioinformatics tools: A qualitative study. *Journal of the Medical Library Association: JMLA*, *95*(4), 454–458. doi:10.3163/1536-5050.95.4.454 PMID:17971896

Wilson, T. D. (1981). On user studies and information needs. *Journal of Librarianship*, *37*(1), 3–15.

KEY TERMS AND DEFINITIONS

Bioinformatics Researchers: A person who is conducting research in the areas of bioinformatics.

Information Needs: A gap in person knowledge that gives the feeling in craving for information either consciously or unconsciously.

Internet: A high speed fiber-optic network global system which used interconnected computer networks through the Internet protocol (TCP/IP) to link several billion devices worldwide.

Journals: A periodical devoted to disseminating original research and commentary on current developments in a specific discipline, sub-discipline, or field of study which is published quarterly, bimonthly and monthly.

Librarian: A person who knows everything but all of which is not applied by him/her. In other words, a librarian is a person who drowns in an ocean but did not swallow the water.

Library: A place where knowledge is store in the form of conventional and non-conventional as well as print and electronic form with efficient professional staffs and users.

Web: A computer programming system to implement literate programming written in a script called Hypertext Mark-up Language (HTML).

Chapter 4
Bioinformatics Database Resources

Icxa Khandelwal
Jaypee University of Information Technology, India

Aditi Sharma
Jaypee University of Information Technology, India

Pavan Kumar Agrawal
G. B. Pant Engineering College, India

Rahul Shrivastava
Jaypee University of Information Technology, India

ABSTRACT

Various biological databases are available online, which are classified based on various criteria for ease of access and use. All such bioinformatics database resources have been discussed in brief in this book chapter. The major focus is on most commonly used biological/bioinformatics databases. The authors provide an overview of the information provided and analysis done by each database, information retrieval system and formats available, along with utility of the database to its users. Most widely used databases have been covered in detail so as to enhance readers' understanding. This chapter will serve as a guide to those who are new to the field of bioinformatics database resources, or wish to have consolidated information on various bioinformatics databases available.

DOI: 10.4018/978-1-5225-1871-6.ch004

INTRODUCTION

The National Center for Biotechnology Information (NCBI) defines bioinformatics as: "the field of science in which biology, computer science, and information technology merge into a single discipline". Bioinformatics can be considered an amalgam of three sub-disciplines:

1. Development of new algorithms as well as statistics so that the relationship between the elements of huge datasets can be determined.
2. Analysis as well as interpretation of biological data i.e. various types of sequences and structures.
3. Development of tools and software to ensure efficient access as well as management of biological data (Toomula, 2011).

The bioinformatics database resources focus primarily on the third sub-discipline of bioinformatics. A database can be defined as a computerized and organized storehouse of related information that provides a standardized way for searching, inserting and updating data. The data stored in these databases is persistent and organized. Database Management System (DBMS) is a software application that deals with the user, other applications, and the database itself in order to perform analysis and capture data in a systematic manner.

Bioinformatics databases or biological databases are storehouses of biological information. They can be defined as libraries containing data collected from scientific experiments, published literature and computational analysis. It provides users an interface to facilitate easy and efficient recording, storing, analyzing and retrieval of biological data through application of computer software. Biological data comes in several different formats like text, sequence data, structure, links, etc. and these needs to be taken into account while creating the databases.

There are various criteria on the basis of which the databases can be classified. On the basis of structure, databases can be classified as a text file, flat file, object-oriented and relational databases. On the basis of information, they can be classified as general and specialized databases. Most commonly, they are classified on the basis of the type of data stored in primary, secondary and composite databases (Kumar, 2005).

CLASSIFICATION OF DATABASES

Type 1

Databases can be classified on the basis of structure as Abstract Syntax Notation (ASN.1), Flat files, Object oriented databases, Relational databases, and XML. Table 1 provides a comparison of various types of databases on the basis of structure

- **ASN.1:** This format comprises of a syntax and description of how a particular data type can be represented physically in a data stream or sequential file (Buneman, Davidson, Hart, Overton, & Wong, 1995). This format has been adopted by NCBI for the representation of sequential data. It is one of the major file formats in GenBank (Cooray, 2012).
- **Flat Files:** This implementation is based on only one table, which incorporates the complete data i.e. all the attributes for each variable. Each row of the table specifies a different record. Specified delimiters are used to differentiate among records. Maintenance of data stored is a major drawback of this type of databases. Integration of two or more databases is difficult due to redundancy in data and variation in the format used.
- **Object Oriented Databases:** Object oriented databases can handle complex data types and can be easily integrated with Object Oriented Programming Languages (OOPL) (Codd, 1970). They can be defined as a collection of objects. Objects represent an instance of an entity and comprise attributes as well as methods (Hasegawa, 2008).
- **Relational Databases:** Relational database systems can be defined as a collection of relations or tables. In a relational database, the data is organized in the form of a table where each row contains a record and each column specific an attribute of the record. The ordering of tuples, attributes or values within a tuple do not make any impact on the relation. The data is subjected to various constraints for validation (Cooray, 2012).
- **XML:** XML can be defined as an advanced flat file format. It provides greater support for representation of complex nested data structures. It contains data definitions and supports new definitions and tags upon requirement. The major advantages of this type are fast accessibility, reliability, and scalability. (Cooray, 2012)

Table 1. Comparison between different databases on the basis of structure

Type	Merits	Demerits	Examples
ASN.1	Implementation ease, Standardized	Not easy to integrate	CDD, GenBank, OMIM
Flat File	User-friendly	Not easy to access, integrate and validate	EMBL, DDBJ
Object Oriented	Implementation ease, Supports abstract data types	Factorization of document, Integration enhancement	MITOMAP
Relational	Implementation ease, Reliable, Scalable	Reduced performance when a large number of join operations are used	GDB, SMART
XML	Fast response, Flexible, Better accessibility	Less mature as compared to DBMS	GO, SwissProt

Cooray, 2012.

Type 2

The databases can be classified into three categories on the basis of the information stored. They are Primary, Secondary and Composite databases.

- **Primary Databases:** Primary databases contain data that is derived experimentally. They usually store information related to the sequences or structures of biological components(Singh, Gupta, Nischal, Khattri, & Nath, 2010)(IASRI, (N.D.)). They can be further divided into protein or nucleotide databases which can be further divided as sequence or structure databases. The most commonly used primary databases are: DNA Data Bank of Japan (DDBJ), European Molecular Biology Laboratory (EMBL) Nucleotide Sequence Database, GenBank, and Protein Data Bank (PDB) (Toomula, 2011).
- **Secondary Databases:** Secondary databases contain the data that is obtained through the analysis or treatment of data present in primary databases. For instance, it can contain conserved protein sequence, signature sequence active site residues of protein families which are obtained from multiple sequence alignment of related proteins, etc. (Varsale, Wadnerkar, Mandage, & Jadhavrao, 2010; Sahoo, Rani, Dikhit, Ansari, & Das, 2009). These databases can be further classified as metabolic pathways database, protein family database, etc. The most common examples are Class Architecture Topology Homology (CATH), Kyoto Encyclopedia of Genes and Genomics (KEGG), Protein Families (Pfam) and Structural Classification of Proteins (SCOP).

- **Composite Databases***:* Composite databases are collections of several (usually more than two) primary database resources. This helps in the lessening the tedious task of searching through multiple databases referring to the same data. The approach used, for instance, the search algorithm employed, differs considerably in every composite database. For example DrugBank offers details on drug and their targets, BioGraph incorporates assorted knowledge of biomedical science and Bio Model is a storehouse of computational models of the biological developments, etc. There are many composite databases which provide users with various tools and software for analysis of data. NCBI being a composite database has stored a lot of sequence of nucleotide and protein within its server and thereby suffers from high redundancy in the data deposited (IASRI, (N.D.).

NATIONAL CENTER FOR BIOTECHNOLOGY INFORMATION (NCBI)

It was developed at the National Institutes of Health (NIH) in 1988 for advancement (Wheeler et al., 2007) of science as well as health as it provides access to a large amount of biomedical and genomic information (www.ncbi.nlm.nih.gov/home/about/mission.shtml). It maintains a large scale of databases and bioinformatics tools as well as services. One of the most popular databases is GenBank. It is a nucleic acid sequence database and its data is acknowledged by the scientists all around the world. Another popular and major database is PubMed, a bibliographic database for the biomedical literature. All the databases are available online at official website of NCBI through the Entrez search engine (Wheeler et al., 2007).

Mission

The aim is to find novel techniques and methodologies for dealing with huge and complex data and provide better accessibility to analytical and computational tools.

Organization

The various branches of NCBI are Computational Biology Branch (CBB), Information Engineering Branch (IEB) and Information Resource Branch (IRB).

Options on Homepage

There are various options can be viewed and explored on the homepage of NCBI's website. They are mentioned in Table 2.

Resources

The resources that are present on this site can be divided into two major categories: databases and tools, which are then further divided as follows:

1. **Databases:**
    a. **General:**
        i. Entrez,
        ii. PubMed and PubMed Central,
        iii. Taxonomy, and
        iv. Protein.
    b. **Gene Level Sequences:**
        i. Gene,
        ii. GenBank,
        iii. Unigene,
        iv. Homologene, and
        v. Reference Sequences.
    c. **Genomic Analysis:**
        i. Entrez Genome.
    d. **Analysis of Gene Expression:**
        i. Gene Expression Omnibus.
    e. **Analysis of Phenotypes:**
        i. Online Mendelian Inheritance in Man, and

Table 2. Various options along with their descriptions present on NCBI's homepage

Option	Description
Submit	Deposition of data
Download	Downloading data from NCBI
Learn	Users can learn about various tools and databases through documents or tutorials
Develop	New applications can be built by using Application Programming Interfaces and code libraries of NCBI
Analyse	Choose an appropriate NCBI tool for a specific data analysis task
Research	Research and collaborative projects of NCBI can be explored

    - ii. Online Mendelian Inheritance in Animals.
  - f. **Molecular Structure and Proteomics:**
    - i. Structure Databases,
    - ii. Molecular Modeling Database, and
    - iii. PubChem.
  - g. Hiv-1/Human Protein Interaction Database.
2. **Tools:**
  - a. BLAST.
  - b. **Gene-Level Analysis:**
    - i. Open Reading Frame Finder.
  - c. **Genomic Analysis:**
    - i. Map Viewer,
    - ii. Model Maker, and
    - iii. Evidence Viewer.
  - d. **Analysis of Gene Expression:**
    - i. Genset, and
    - ii. Probe.
  - e. **Tools Supporting Proteomics Blast Link (Blink):**
    - i. Open Mass Spectrometry Search Algorithm (Wheeler et al., 2007).

Entrez Global Query Cross-Database Search System

It is an integrated database search and retrieval system which is widely used (Maglott D., 2005) as it enables text searching using Boolean expressions (Schäffer et al., 2001). The data is integrated from a wide variety of sources and databases to create a uniform information model. Hence it is used for both indexing as well as retrieval (H., 2014). It aids in the availability of extensive links within and between database records (Wheeler et al., 2007). It allows users the combined access to sequential, structural and taxonomic data. Graphical representation of chromosome maps as well as the sequence is also provided. It can also aid the user in obtaining the sequences, structures or references that are related to the query entered. The users can also store their private configuration options using 'My NCBI' (Wheeler et al., 2007).

PubMed and PubMed Central

NCBI National Library of Medicine (NLM) created the PubMed database which is also the part of NCBI Entrez retrieval system. Providing users with an ease in accessing abstracts as well as references from biomedical and life sciences journals was the primary reason behind its creation. To add on, there are links provided for accessing the complete journal articles (Lindberg, 2000). The primary data source for PubMed is MEDLINE database (C. K. J. J. C. M. J., 2000). PubMed Central (PMC) enables users to access freely all the articles it contains (B. J. S. E., 2002).

Taxonomy

Taxonomy database can be accessed using the Taxonomy Browser for either viewing the taxonomic position or retrieving data depending upon the requirement of the user (Wheeler et al., 2007).

Protein

This database stores individual protein sequences in a textual format including FASTA and XML. The most common sources from which these sequences have been obtained are GenBank, NCBI Reference Sequence (RefSeq) project, PDB and SWISS-Prot/UniProtKB. Sets of similar and identical proteins determined by BLAST is also provided for each sequence (S. E., 2013).

GenBank

It is located in the USA. NCBI since 1992 has provided access to GenBank DNA sequence database through NCBI gateway server and hence is accessible freely(Schuler, Epstein, Ohkawa, & Kans, 1996). The three nucleotide sequence databases GenBank, European Molecular Biology Laboratory (EMBL) and DNA Data Bank of Japan (DDBJ) coordinate among themselves so that all three of them are updated with the latest findings (Pruitt, Tatusova, & Maglott, 2005).

A detailed structure of a nucleotide sequence file format in this database includes the following:

1. **Locus:** This can be defined as a title given by GenBank itself to name the sequence entry. It includes the following:
    a. **Locus Name:** Similar to accession number for the sequence.
    b. **Sequence Length:** Tells the number of bases existing in the sequence.
    c. **Molecule-Type:** Identifies the type of nucleic acid sequence. The various types are mRNA (which is present as cDNA), rRNA, snRNA, and DNA.
    d. **GB Division:** Postulates class of the data according to classification criteria of GenBank.
    e. **Modification Date:** The date on which the record was modified.
2. **Definition:** This denotes the name of the nucleotide sequence.
3. **Accession:** This covers accession number, accession version, and GI number. Accession number can be defined as the unique identifier associated with each

nucleotide sequence present in the database. If more than one record is created for a particular sequence then it will have the same accession number but all records will have different versions associated with that accession number.

4. **Keyword:** Defined words that were used to index the entries.
5. **The Source:** This describes organism from which sequences have been obtained. The accepted common name is mentioned first and then the scientific name is mentioned. In the end, the taxonomic lineage according to GenBank is specified.
6. **The Citation:** Includes the journal from which with the sequence was derived as initially the sequences were obtained only from published literature.
7. **Features:** These consist of the information derived from the sequence such as biological source, coding region, exon, intron, promoters, alternate splice patterns, mutations, etc.
8. **Sequence:** Contains the following:
    a. Count of presence of each nucleotide in the sequence,
    b. Whole nucleotide sequence,
    c. Beginning of sequence is determined by keyword "ORIGIN", and
    d. End is marked as "\\".

There are many techniques for retrieving and searching data from GenBank. The sequence identifiers can be searched in GenBank along with Entrez Nucleotide. Another approach is using BLAST search and then aligning nucleotide sequences to the query sequence. The last method is to search the appropriate link and then download nucleotide sequences. It intends to offer and reassure access of the most updated data of the nucleotide (www.ncbi.nlm.nih.gov/genbank/).

In order to maintain the confidentiality, GenBank on request, reserves announcement of new submissions for a definite interval of time. If sequences of the human genome are deposited to GenBank, it is mandatory not to include any personal information that can anyhow lead to the revelation of the identity of the individual.

Gene

It is involved in the characterization and organization of gene information about genes. Each gene record can be identified through a unique GeneID (Maglott D., 2005). Organism-specific XML files can be created by applying the organism filter (Maglott, Ostell, Pruitt, & Tatusova, 2007).

UniGene

UniGene can be defined as a software that partitions the sequences present in GenBank into sets of non-redundant gene- oriented clusters. These collections have been employed in the creation of unique sequences for microarray fabrication in order to comprehend gene expression study on a large scale (Schuler, 1997).

Entrez Genome

Entrez Genome (Tatusova, Karsch-Mizrachi, & Ostell, 1999) provides access to a large number of complete microbial genomic sequences and an enormous number of viral genomic sequences. There are also a huge amount of reference sequences for eukaryotic organelles. It is supplemented by Entrez Genome Project database. This database provides the status of various on-going annotation, assembly and sequencing projects (Wheeler et al., 2007).

Gene Expression Omnibus (GEO)

It can be defined as a data repository as well as retrieval system for various types of high-throughput molecular abundance data. It accepts array comparative genomic hybridization (aCGH) data, chromatin immune- precipitation on array (ChIP-chip) data, gene expression data and SNP array data (Barrett et al., 2007).

Online Mendelian Inheritance in Man (OMIM)

It contains a catalog of human genes as well as genetic disorders (Hamosh, Scott, Amberger, Bocchini, & McKusick, 2005). The data comprises of disease phenotypes, gene polymorphism, genes, map locations and patterns of inheritance (Wheeler et al., 2007).

Online Mendelian Inheritance in Animals (OMIA)

It is a database that contains the genes, inherited disorders as well as traits in fauna species excluding human and mouse. It also has the links to the records that are relevant in OMIM, PubMed and Gene databases (Wheeler et al., 2007).

PubChem

PubChem can be considered as a molecular library containing relational databases which were created using Microsoft SQL servers. The major focus is on the biological, chemical and structural properties of small molecules so that they can be used as diagnostic and therapeutic agents. All of the deposited data can be accessed freely by the users. It comprises of three sub-databases which are: PCSubstance, PCCompound, and PCBioAssay which contain substance information, compound structures and bioactivity data of compounds respectively (Wheeler et al., 2007).

Basic Alignment Local Search Tool (BLAST)

NCBI developed BLAST, a powerful tool for comparing sequences from various organisms (T., 2002). It can be defined as an algorithm to determine the similarity between biological sequences (Altschul, Gish, Miller, Myers, & Lipman, 1990). Gapped alignments having links to the database are provided in the final result (Wheeler et al., 2007). It has been reported that this tool has the capacity to search entire DNA database in less than 15 seconds. (ftp://ftp.ncbi.nlm.nih.gov/pub/factsheets/HowTo_BLASTGuide.pdf). The input is in the FASTA or Genbank format and output can be displayed in various formats like HTML, XML formatting and plain text (T., 2002). The score of each alignment is assigned an Expectation Value (E-value) which is a measure of statistical significance (Wheeler et al., 2007).

Open Reading Frame Finder

Six-frame translation of nucleotide sequence is performed by this tool. The result is location of each ORF within a specified size range (Wheeler et al., 2007).

EUROPEAN MOLECULAR BIOLOGY LABORATORY (EMBL)

It is a molecule-based biology research foundation, maintained by 21 member states and formed in the year 1974 (www.embl.fr/aboutus/general_information/organisation/member_states/index.html). It stores and makes available raw nucleotide sequences. It is situated in UK (IASRI, (N.D.)). European Bioinformatics Institute (EBI) maintains EMBL nucleotide sequence database (Garg, Pundhir, Prakash, & Kumar, 2008).

Mission of EBI

The various aims of the organization are as follows:

- To provide freely available data and bioinformatics services to all facets of the scientific community.

- To contribute to the advancement of biology through basic investigator-driven research.
- To provide advanced bioinformatics training to scientists at all levels.
- To help disseminate cutting-edge technologies to industry.
- To coordinate biological data provision throughout Europe (www.ebi.ac.uk/).

European Nucleotide Archive (ENA)

Free as well as unrestricted information access on DNA and RNA sequences is provided by ENA. This archive is created using three databases which are Sequence Read Archive, Trace Archive and EMBL Nucleotide Sequence Database (www.ebi.ac.uk/ena). The information in ENA can be extracted manually or programmatically and resultant files can be obtained in various formats like XML, HTML, FASTA and FASTQ. (O. J., 2002) Using accession numbers and other specific text queries the users can obtain individual archives (Leinonen et al., 2010).

EMBL Nucleotide Sequence Database

It contains the high level genome assembly data of sequences and their functional annotation (Stoesser et al., 2003; Amid et al., 2011). The data is store in flat file format.

Data Classes

The different data classes of sequences are mentioned in Table 3.

Resources at EMBL-EBI

The EBI website provides link to access many services like various biological tools and databases. Some of the most common resources are listed below in the Appendix.

DNA DATA BANK OF JAPAN (DDBJ)

This biological database resource belongs to National Institute of Genetics (NIG) in Japan. DDBJ is the only nucleotide sequence data bank currently present in Asia. Although DDBJ essentially has Japanese researchers as contributors but it also accepts the data from researchers of other countries. It is an associate of the International Nucleotide Sequence Database Collaboration (INSDC). The major driving force behind DDBJ operations is the advancement of the quality of INSD

Table 3. Summary of data classes

Data Class	Definition	Example
EST	Raw expressed sequence tags without sequence quality information	FASTA Flat file HTML XML
WGS	Genomic contigs	FASTA Flat file HTML XML
GSS	Genome survey sequence; single pass, single direction sequence	FASTA Flat file HTML XML
HTC	High throughput assembled transcriptomic sequence and optional annotation	FASTA Flat file HTML XML
HTG	High throughput assembled genomic sequence and optional annotation	FASTA Flat file HTML XML
STD	Assembled and annotated sequences	FASTA Flat file HTML XML
CON	Scaffolds build from genomic or transcriptomic contigs	FASTA Flat file HTML XML
STS	Sequence tagged site	FASTA Flat file HTML XML
PAT	Patent sequences	FASTA Flat file HTML XML
TSA	Transcriptomic contigs	FASTA Flat file HTML XML
CDS	Coding sequences	FASTA Flat file HTML XML

www.ebi.ac.uk/ena/submit/sequence-format.

as the nucleotide sequence accounts organism development more directly than other biological constituents.

Tasks

Key tasks of DDBJ Center are as follows:

1. Construction and operation of INSDC which offers nucleotide and amino acid sequence data along with the patent request.
2. Provides searching and analysis of biological data.
3. Training course and journal.

DDBJ Flat File Format

The data submitted in DDBJ is managed and retrieved according to the DDBJ format (flat file). The flat file includes the sequence and the information of who submitted the data, references, source organisms, and information about the feature, etc (www.ddbj.nig.ac.jp/ddbjingtop-e.html).

PROTEIN DATA BANK (PDB)

PDB is a universal free archive for structural data of biological macromolecules. It was established in 1971 at Brookhaven National Laboratory under the governance of Walter Hamilton and initially contained only 7 protein structures. There were two major reasons which initiated the formation of PDB. The first one was increasing assembly of datasets related to protein structure. Another reason was the availability of Brookhaven Raster Display (BRAD) that envisages the protein structures in 3-D (Meyer, 1997). It is now conserved by the Research Collaboratory for Structural Bioinformatics (RCSB). It contains three-dimensional structures of proteins, nucleic acid fragments, RNA molecules, large peptides and complex structures of proteins and nucleic acids (Berman et al., 2000).

Data Storage and Acquisition

The data for each structure is stored in a distinct file and hence the data is stored in flat file arrangement. The major source for the three-dimensional structure of proteins includes cryo-electron microscopy, molecular modeling, NMR experimentations and X-ray crystallography trials (IASRI, (N.D.)). Each structure present within PDB is given four character alphanumeric characters. The protein structure files may be viewed using the open resource. The RCSB PDB website covers an extensive list of both free and marketable molecule conception programs that includes Jmol, Pymol, and Rasmol and web browser plugins (Protein_Data_Bank, N.D.).

Atomic Coordinate Entry Format Description

The various sections of the PDB file are:

1. Title Section,
2. Primary Structure Section,
3. Heterogen Section,
4. Secondary Structure Section,
5. Connectivity Annotation Section,
6. Miscellaneous Features Section,
7. Crystallographic and Coordinate Transformation Section,
8. Coordinate Section,
9. Connectivity Section, and
10. Bookkeeping Section.

Title Section

It contains the elements which describe the experiment and biological macromolecules present in the record entered. The elements of this section along with their description are mentioned below:

- **HEADER:** It contains an idCode field which is used for unique identification, a classification, and date when the coordinates were deposited to the PDB archive for a PDB record.
- **OBSLTE:** This element appears in records which have now been removed. Also, the new entries which have replaced those records are also displayed.
- **TITLE:** It stores the information of title of experiment or analysis corresponding to the record.
- **SPLIT:** If a specific entry comprises a portion of the huge macromolecular complex then using this element all the PDB records belonging to that complex can be identified.
- **CAVEAT:** The users are warned about errors and unresolved issues in the record.
- **COMPND:** It is used to describe the macromolecular contents of a record.
- **SOURCE:** Biological and/or chemical source are specified.

- **KEYWDS:** A set of relevant terms which can be further used to categorize the record are provided.
- **EXPDTA:** It tells the experimental technique employed for structure identification.
- **NUMMDL:** It provides the total number of models in a PDB record.
- **MDLTYP:** It comprises of additional annotation.
- **AUTHOR:** It has the names of people who are responsible for contents of a particular record.
- **REVDAT:** It stores the details of modifications made to a record since it was released.
- **SPRSDE:** It stores a list of record ID codes which were made obsolete by a given coordinate record.
- **JRNL:** It has the primary literature citation which contains the description of the experiment that resulted in the deposited coordinate set.
- **REMARK:** It contains annotations, comments, experimental details and information not included in other elements (Berman et al., 2003).

Primary Structure Section

It contains the sequence of residues present in various chains of the macromolecules. The elements of this section along with their description are mentioned below:

- **DBREF (Standard Format):** It is used to provide cross-reference links between PDB sequences and a corresponding database sequence.
- **DBREF1/DBREF2:** It is used when the accession code or sequence numbering does not fit the space allotted in the above format.
- **SEQADV:** It is used for the identification of differences between sequence information present in SEQRES and DBREF.
- **SEQRES:** It has the list of consecutive chemical components which are covalently linearly in order to form a polymer.
- **MODRES:** It contains the description of modifications made in the protein and nucleic acid residues (Berman et al., 2003).

Heterogen Section

It has a complete description of non-standard residues in a PDB record. The elements of this section along with their description are mentioned below:

- **HET:** It provides a description of non-standard residues.

- **HETNAM:** It assigns the chemical name of the compound with the given hetID.
- **HETSYN:** It tells synonyms for a compound in the corresponding HETNAM element.
- **FORMUL:** It presents chemical formula and charge of a non-standard group (Berman et al., 2003).

Secondary Structure Section

It contains description of helices and sheets found in structures of protein as well as polypeptide. The elements of this section along with their description are mentioned below:

- **HELIX:** It identifies the position of helices in the molecule.
- **SHEET:** It identifies the position of sheets in the molecule (H. Berman et al., 2003).

Connectivity Annotation Section

It contains the description of existence and location of disulfide bonds and other linkages. The elements of this section along with their description are mentioned below:

- **SSBOND:** It provides the identification of each disulfide bond in protein and polypeptide structures.
- **LINK:** It provides a specification of connectivity between residues.
- **CISPEP:** It provides a specification for prolines and other peptides found in cis conformation (H. Berman et al., 2003).

Miscellaneous Features Section

It describes the properties of the molecule. Element of this section along with its description is mentioned below:

- **SITE:** It specifies the different types of residues present in the structure (Berman et al., 2003).

Crystallographic and Coordinate Transformation Section

It contains the description of the geometry of the crystallographic experiment and the coordinate system transformations. The elements of this section along with their description are mentioned below:

- **CRYST1:** It stores the unit cell parameters.
- **ORIGXn (where n = 1, 2, or 3):** It stores transformation from orthogonal coordinates to submitted coordinates.
- **SCALEn (where n = 1, 2, or 3):** It stores transformation from orthogonal coordinates to fractional crystallographic coordinates.
- **MTRIXn (where n = 1, 2, or 3):** It stores transformations expressing non-crystallographic symmetry (Berman et al., 2003).

Coordinate Section

It stores a collection of atomic coordinates as well as the MODEL and ENDMDL records. The elements of this section along with their description are mentioned below:

- **MODEL:** It provides a specification of the model serial number.
- **ATOM:** It provides atomic coordinates for standard amino acids and nucleotides.
- **ANISOU:** It has anisotropic temperature factors.
- **TER:** It determines the end of a list of ATOM/HETATM records for a chain.
- **HETATM:** It is used for the representation of non-polymer or other "non-standard" chemical coordinates.
- **ENDMDL:** It is paired with a corresponding MODEL element to generate individual structures (Berman et al., 2003).

Connectivity Section

It contains information on atomic connectivity. Element of this section along with its description is mentioned below:

- **CONECT:** It determines connectivity between atoms for which coordinates have been supplied (Berman et al., 2003).

Bookkeeping Section

It has final information about the file itself. Element of this section along with its description is mentioned below:

- **MASTER:** It contains the number of lines in the coordinate file for selected record types.
- **END:** It denotes the end of the PDB file (Berman et al., 2003).

UNIPROT

It is a database of freely accessible protein sequences which contains high-quality data and functional information for the proteins. Many of the records have been obtained from genome sequencing projects. The information regarding the biological function of the protein has been extracted from the research literature. European Bioinformatics Institute (EBI), Swiss Institute of Bioinformatics (SIB) and Protein Information Resource (PIR) constitute the UniProt consortium. Each one of them is deeply engaged in protein database maintenance and annotation (Consortium, 2011)(Eck & Dayhoff, 1966). It includes four core databases: UniProtKB, UniParc, UniRef, and UniMes. It is funded by grants from European Commission, National Human Genome Research Institute, NIH, NCI-caBIG, the Department of Defense and Swiss Federal Government (Consortium, 2010).

UniProtKB

UniProt Knowledgebase (UniProtKB) is a protein database that is partially curated by experts. It includes three databases: Swiss-Prot, TrEMBL, and PIR-PSD. The former one contains reviewed and manually annotated records whereas the latter one comprises the un-reviewed and automatically annotated entries (Consortium, 2010).

Swiss-Prot

It can be defined as a manually curated protein sequence database with high annotation level. It was created by Amos Bairoch in 1986, developed by the Swiss Institute of Bioinformatics and subsequently further enhanced by Rolf Apweiler at EBI (Bairoch 2000; Bairoch & Apweiler, 1996; Altairac, 2006). It is well known for its high annotation level, a low degree of redundancy, standardized nomenclature usage, and links to specialized databases (O'Donovan et al., 2002).

To provide a set of relevant information for a particular protein is the core aim of this database. It also aggregates data obtained from scientific literature and bio curator-evaluated computational analysis. To remain up-to-date the annotations are often reviewed periodically (Apweiler, Bairoch, Wu, et al., 2004). It has Sequence Retrieval System (SRS) that helps in searching through relevant databases like for Translated EMBL (TrEMBL) on the same site.

Each data in SwissProt that belongs to a protein sequence is considered to have a separate core data and annotation. Former includes protein sequences, related references, bibliography and taxonomy of the organism from where the sequence has been extracted. Latter includes function(s) performed by the protein, post-translational modification, functional sites, structural domain sites, secondary structure features, likenesses to other proteins and diseases that may be caused due to a mutation in diverse strains (IASRI, (N.D.)).

Table 4 describes the codes which are used in a record of Swiss-Prot.

TrEMBL

Since the sequence data were being generated at a very high rate with respect to the ability of the SwissProt in performing the annotation hence TrEMBL Nucleotide Sequence Data Library was created so as to facilitate computer-annotated data for those proteins which could not be entered in Swiss-Prot (Apweiler, Bairoch, & Wu, 2004). The records present in this database are automatically annotated and have been analyzed computationally with high quality. Automatic processing and insertion of the translation of annotated coding sequences present in the three major nucleotide sequence database are done through this database. It also takes into account the sequences from PDB, Ensembl, RefSeq and CCDS (Consortium, 2011).

PIR: International Protein Sequence Database (PIR - PSD)

PIR was developed initially at National Biomedical Research Foundation (NBRF) in the year 1984. PIR, Munich Information Centre for Protein Sequences (MIPS) and Japan International Protein Information Database (JIPID) collaborated together to form PIR-PSD(IASRI, (N.D.)) which can be defined as an integrated public bioinformatics resource to support genomic and proteomic research. It also helps in the identification and interpretation of protein sequences. Since 2002, it has become a part of UniProtKB (Wu & Nebert, 2004). This database contains non – redundant data, is annotated by experts, is comprehensive in nature and uses object-oriented DBMS (Database Management System). Classification of the sequences of protein on the basis of super-family is the unique characteristic of this database. The clas-

Table 4. Codes along with the full-name and description as used in Swiss-Prot

Code	Full- Name	Description
ID	Identification	It is a unique identifier that related with each entry and it appears in the beginning of a record.
AC	Accession Numbers	The name of data record might change but the AC cannot be changed. Also, if there are more than one accession numbers it suggests that the record was fabricated by integration with other records.
DT	Date	It comprises date consistent with the data entry formation, alteration date of sequence and annotation respectively.
DE	Description	It contains the general details of the sequence.
GN	Gene name	It contains the name of the encoding gene(s).
OS, OG, OC	Organism name, organelle, organism classification	These have the details of name and taxonomy of the organism.
RN, RP, RX, RA, RT, RL	Reference number, Position, comments, cross-reference, authors, title, and location	These have the bibliographic information.
CC	Comments	It has the free text comments.
DR	Database cross reference	It provides cross-reference to other databases.
KW	Keywords	It contains a list of keywords that can be used in indexes.
FT	Features tables	It defines regions or sites of concern in the sequence.
SQ	Sequence headers	It directs beginning of sequence statistics and gives a short summary of its contents.

IASRI, (N.D.).

sification criteria also takes into account the homology domain as well as sequence motifs (IASRI, (N.D.)).

The curation status of PIRSF Database can be categorized as uncurated, preliminary and full/Full (with description). PIRSF Membership can be classified as follows: full (F), associate (A) and seed (S). Full is used for proteins which share complete sequence similarity as well as common domain architecture. Associate is used for those members whose sizes are greater than the family length range. Seed is used when it is required to create family specific full-length and domain HMMs using already present full members. PIRSF Family Level are divided into the Homeomorphic family (HFam), Subfamily (SubFam) and Superfamily (SuperFam).

The members which are categorized as HFam family are homologous as well as homeomorphic. The category of subfamily delineates protein clusters within a homeomorphic family which have specialized functions and/or variable domain architecture. Superfamily level brings together a number of distantly related families and orphan proteins that share one or more domains (http://pir.georgetown.edu/pirwww/support/help.shtml).

There are various rules for writing a protein pattern and they are as follows:

1. Usually, capital letters are used for denoting amino acid residues and "-" is inserted in between two amino acids.
2. To provide multiple amino acids as a choice in a particular position "[...]" can be used.
3. To exclude a set of amino acids "{...}" is used.
4. If a particular position has "x" then it means that any amino acid can be inserted there.
5. If an amino acid occurs more than once consecutively then this can be denoted using "(n)", where n is a number of times that amino acid has occurred.
6. On similar grounds, as the previous rule, "(n1,n2)" are employed for multiple or variable positions.
7. To match a pattern at N or C terminus, the symbol ">" is required at the beginning or end respectively.

UniParc

UniProt Archive (UniParc) stores proteins sequences from publicly available protein sequence database in a non-redundant manner (Apweiler, Bairoch, Wu, et al., 2004) and it is updated on a regular basis. Since proteins may exist in several databases and there are high chances that a single sequence is present multiple times in the same database. Hence to avoid redundancy of data, each unique sequence is presented only once in this database. The identical sequences are merged even if they belong to different species. A unique identifier called UPI is given to each sequence which enables the identification of a same protein from various source databases. The protein sequences present in this database are without any annotation. Database cross-references are provided in order to facilitate the retrieval of more detailed information from the source databases.

UniRef

Clustered sets of protein sequences from UniProtKB and selected UniParc records are comprised in UniProt Reference Clusters (UniRef).(Suzek, Huang, McGarvey, Mazumder, & Wu, 2007) The UniRef100 database is involved in combining the identical sequences and sequence fragments (from any species) into a single UniRef entry. The accession numbers of all the merged records, sequence of a representative protein and links to the corresponding UniProtKB and UniParc records are mentioned. (Consortium, 2010)

UniMES

UniProt Metagenomic and Environmental Sequences (UniMES) database has been created for environmental and metagenomic data (Consortium, 2010). In order to improve the original data through more analysis, proteins that have already been predicted are merged using InterPro. UniMES is the source containing data from Global Ocean Sampling Expedition (GOS) (Yooseph et al., 2007). UniProt Reference Clusters or UniProt Knowledgebase do not contain data of environmental sample of this database.

STRUCTURAL CLASSIFICATION OF PROTEINS (SCOP)

SCOP database was started by Alexey Murzin in the year 1994 at Centre of Protein Engineering and was later on established at Laboratory of Molecular Biology (Chandonia et al., 2007) in Cambridge University, England (Hubbard, Ailey, Brenner, Murzin, & Chothia, 1999). It is an open source database and hence is accessible freely (Subramanian, Muthurajan, & Ayyanar, 2008). The structural domains of proteins were manually classified using the criteria of likeliness of sequences as well as structures in order to create this database (Conte et al., 2000; Andreeva et al., 2004). Its foremost function is to categorize 3D structures of proteins in a hierarchical pattern of structural levels which include family, super-family, fold and class so as to determine the evolutionary association between proteins. (IASRI, (N.D.))

SCOP Levels

PDB contains freely available 3D protein structures which are used by SCOP for classification. The protein domain is the main element of classification of structure in SCOP (Hubbard et al., 1999). The various levels are described in Table 5.

Table 5. Different levels of SCOP database

Level	Description
Class	Types of folds.
Fold	Various domain shapes contained in one single class.
Superfamily	Domains having at-least a distant common ancestor and belonging to a particular fold are clustered together.
Family	Domains sharing a more recent common ancestor in a super-family are grouped into families.
Protein Domain	Clustered domains in families.
Species	Gathering of protein domains in accordance with species.
Domain	Either an entire protein or a part of protein.

Hubbard et al., 1999.

The root of this hierarchical classification is class which has the following major types:

1. All α proteins class which contains domains comprising α-helices.
2. All β proteins class which contains domains comprising β-sheets.
3. α and β proteins class which contains individual units as β-α-β and β-sheets are usually parallel to each other.
4. α or β proteins class which contains separated α and β regions and β-sheets are arranged in an anti-parallel fashion.
5. Multi-domain proteins class which contains those folds comprise at-least two domains of different classes.
6. Membrane and cell surface proteins as well as peptides class which contains proteins excluding the ones present in the immune system.
7. Small proteins class which contains proteins that have a metal ligand, heme and/or disulfide bridges.

It can be inferred from the above table that families are more closely related than super-families. The criteria for placing various domains within a fold into the same family is that either they share a minimum of 30% sequence similarity or they share the same function with 15% sequence similarity. To support the classification of domains into super-families and families BLAST is used (Andreeva et al., 2004; Hubbard et al., 1999).

Representation

The different types of each level are represented using a Sunid. It is a unique identifier for each node in the SCOP hierarchy. Each fold is supplemented by an explanation of that fold (Hubbard et al., 1999). For instance, Folds belonging to a particular class will be represented as the following:

1. **Name of Fold-Type 1 [SUNID] (No. of Elements in this Group):**
    a. Description of fold-type 1.
2. **Name of Fold-Type 2 [SUNID] (No. of Elements):**
    a. Description of fold-type 2.

SCOP Concise Classification String (SCCS) can be used to represent the families in SCOP which has the format as Alphabet.Number.Number.Number where alphabet determines class and numbers refer to fold, superfamily, and family, respectively (Conte, Brenner, Hubbard, Chothia, & Murzin, 2002).

SCOP Successors

SCOPe stands for Structural Classification of Proteins extended, a database which was made public in the year 2012. Also, in the year 2014 manual curation was again established into the database SCOPe in order to retain unambiguous protein structure assignment.SCOP2 is another system for classification with inclination more towards the complexity of evolution which is an innate property of every protein (http://scop.berkeley.edu/help/ver=2.05#scopchanges).

CLASS, ARCHITECTURE, TOPOLOGY, AND HOMOLOGOUS SUPERFAMILY (CATH)

CATH, a database for hierarchical classification of protein domains was developed at University of London (IASRI, (N.D.)). It is known to be semi-automatic in nature and comprises tools that help in comparison of the general structure of protein. Both automatic as well as manual approaches are intricate in protein classification (Orengo et al., 1997).

Employing the above-mentioned classification approach, identification of protein domains structures is done and thereby similarity and dissimilarity are noted down in order to discern homology relationships and other structural similarities. When there is enough evidence to prove that given protein domains have diverged from the same family then they are group into super-families.

Applying Hidden Markov search software named as HMMER3, use of an in-house algorithm known as DomainFinder for the purpose of reconciling the potential matches into a unified multi-domain architecture abbreviated to Mama process for modeling protein sequence discrepancy within domain super-families is worked upon with the aim of reaching to major proteins sequence repositories. A resource Gene3D helps in the presentation of these predicted sequence domains. Also, once or twice in year latest updates are provided to the database. There is the likelihood that PDB record required by a user has not been curated till date. Hence, in order to overpower this issue, the user can either search the structure using CATH domain recognition algorithm called as CATHEDRAL or can go ahead with the pre-released information (Sillitoe et al., 2015).

Classification Levels

The various classification levels of this database are given in Table 6. The first four levels have been manually curated and remaining ones are classified through automatic sequence clustering protocol.

Comparison of CATH and SCOP

CATH and SCOP share many similarities but there are many areas where their detailed classification varies significantly, as seen in Table 7.

Classes in CATH

CATH defines four classes which are: mostly-α, mostly-β, α and β as well as few secondary structures.

Methodology of CATH

The methodology adopted by this database is as follows:

1. Separation of proteins into various domains
2. Created domains are then automatically sorted into classes.
3. Clustering is performed on the basis of sequence similarity upon the created classes.
4. Groups generated in step 3 constitute the H levels of the classification.
5. Topology level is created by structural comparison of H levels.
6. Architecture level is not assigned computationally.

Table 6. Different classification levels of CATH database

Level	Representing Letter	Level Name	Criteria for Classification
1	C	Class	Content of secondary structure
2	A	Architecture	General spatial arrangement of secondary structures
3	T	Topology	Spatial arrangement and connectivity of secondary structures (fold)
4	H	Homologous Superfamily	Evidence of evolutionary relationship shall be obtained through manual curation with similarity in at least two criteria out of three i.e. sequence, structure or function.
5	S	Sequence Family	Sequence similarity is >=35%
6	O	Orthologous Family	Sequence similarity is >=60%
7	L	"Like" domain	Sequence similarity is >=95%
8	I	Identical domain	Sequence similarity is 100%
9	D	Domain counter	Unique domains

Sillitoe et al., 2015.

Table 7. Analogy between the levels of SCOP and CATH

Serial Number	CATH	SCOP
1	Class	Class
2	Architecture	Fold
3	Topology	
4	Homologous superfamily	Super-family

Hadley & Jones, 1999; Day et al., 2003.

7. At the end, Class Level classification is performed based on the following 4 criteria:
   a. Content of secondary structure,
   b. Contacts in secondary structure,
   c. Alteration scores of secondary structure, and
   d. Parallel strands percentage(Sillitoe et al., 2015).

Search Options Provided by CATH

The database can be searched by one of the following ways:

1. Entering a text, ID or keyword,
2. Uploading a protein sequence in FASTA format,
3. Uploading a PDB structure, and
4. Browsing the hierarchy (Sillitoe et al., 2015).

Data Files Provided by CATH

The various types of data files provided by CATH along with their description are mentioned in Table 8.

Using the Online Database

The classic investigation in this database begins by tracing a single PDB structure to its function and homologues. Let us assume that a user is interested in domains

Table 8. Type and description of data files provided by CATH

Type of File	Description of File
CathCathedral library	It is the library containing graphs of secondary structure of domain.
CathDomainDescriptionFile	It contains complete information of various domains in CATH.
CathDomainList	It contains a list of CATH domains which have been assigned already.
CathDomainPdb	It is a library containing PDB files which have been chopped for representative CATH domains.
CathDomall	It stores for each PDB Chain domain boundaries in "domall" format.
CathHmm	It contains HMM library file for CATH domains.
CathHmm (+unclassified)	It contains HMM library file for all CATH domains, including those which have not been classified till now.
CathNames	It contains a list of manually assigned names of CATH classification nodes.
CathUnclassifiedList	It contains a list of CATH domains that have not been classified till now.
Chains	It contains a list of PDB chains in CATH.
Domain Sequences (ATOM)	It is a FASTA sequence database created through ATOM records in PDB for all CATH domains.
Domain Sequences (COMBS)	It is a FASTA sequence database created through COMBS sequence data for all CATH domains.
Representative Domain Sequences (ATOM)	It is a clustered FASTA sequence database for CATH domains.
Representative Domain Sequences (COMBS)	It is a clustered FASTA sequence database for CATH domains.
Representatives	It has a list of CATH representative domains.

Sillitoe et al., 2015.

that can be found in a particular PDB chain. The user has the PDBID of the desired protein. The steps to be performed are as follows:

1. In the top right corner of home page there are links to key functions performed and the 'Quick Search' box. The user can enter the PDBID into this box and then click on Enter button.
2. The result page then obtained will have a list of all records that match with the query –term. Chain, Domains, Node and PDB are the return types obtained.
3. The domain in which user is interested will have a CATH Domain code which can be defined as a PDB Chain Identifier extension.
4. If on results page, user clicks on PDB record then structures and sequences for the chains corresponding to that record can be viewed. There is also a tabular representation of corresponding chains and domains.
5. When the user goes to the pages corresponding to the query domain then it will be observe that a tab entitled 'History' has been inserted. This tab contains the actions taken by curators of this database for assignment of domains.
6. Domain recognition in structures is done using CATHEDRAL.

SSAP is a server provided by this database for pair-wise comparison of two structures (Sillitoe et al., 2015).

PFAM

A database of protein families, Pfam contains annotations as well as multiple sequence alignments generated using hidden Markov models (Finn et al., 2009; Finn et al., 2006; Bateman et al., 2004). There are 4 elements present in all the families or patterns. These are: annotation, seed alignment, HMM profile and full alignment of sequences. Using seed alignment, sequences are bootstrapped into multiple alignments and eventually family (IASRI, (N.D.)).

Features

The user can get the information about known protein structures, multiple alignments, protein domain architectures and species distribution for each family in Pfam. It has been reported that Pfam contains minimum one match for 80% of protein sequences present in UniProt Knowledgebase. Now, the Pfam consortium is involved in the coordination of the annotation of Pfam families via Wikipedia (Finn et al., 2009).

Types

The two sub-types of Pfam database are Pfam-A and Pfam-B. The former one is manually curated. It stores protein sequence alignment and hidden Markov model for each record. Since the records in Pfam-A are unable to consider all known proteins, Pfam-B was created to serve as an automatically generated supplement. A huge number of small protein families obtained from clusters are present in Pfam-B(Heger, Wilton, Sivakumar, & Holm, 2005). When no Pfam-A families are found then Pfam-B families were used earlier but it has been discontinued from release 28.0 (Finn et al., 2009). iPfam (Finn, Marshall, & Bateman, 2005) is a database that was built on domain description provided by Pfam. It determines whether the different proteins described together in PDB are so enough to potentially interact.

Pfam 28.0

This is the latest version of Pfam. This database creates the higher-level clustering of families which are related and these groupings are termed as clans. A clan can be defined as a collection of Pfam-A records that which are related to each other by sequence, structure or profile-HMM similarity. A family in Pfam is now usually retrieved to as a Pfam-A record. Each record includes seed alignment which is curated, profile HMMs and full alignment which has been automatically generated (Finn et al., 2009).

Classification of Pfam Records

Each record in Pfam can be classified into one of four ways which are: Family, Domain, Repeat and Motifs. Family can be defined as a collection of related protein regions. Domain is a structural unit of a protein that can evolve and function independently. Repeat is a term used for a short unit which becomes stable when there are multiple copies but in isolation it is unstable. Motif is a short unit which is present outside globular domains (Finn et al., 2009).

Pfam-A Family Page

This is a page through which a user is able to view Pfam annotation for a protein family. Also, domain architectures can be sighted where information regarding a particular family is developed; it supports alignments in several formats for the family, therefore can be downloaded. To add on, for each family information such as structural details, phylogenetic, the HMM logo and species distribution are made accessible.

Clan

A clan can be defined as a category of protein families that are triggered from the single evolutionary origin. The likeness in tertiary structure or common motif sequence of the protein indicates the evolutionary relationships. Clan alignment is the alignment of the seed alignments of all the families within a particular clan.

Criteria for Categorizing Families into Clans

There are a several criteria used for classifying families into clans. The golden standard is the usage of structures for making the classification. Profile comparisons such as HHsearch are used in the absence of structures. Sequence that matches two HMMs in the identical region of protein sequence is considered. SCOOP is a method that takes into consideration the common matches when searched, which may thereby specify an association. Such type of information is employed in order to decide about relationship among families.

Site Organization

The major page for accessing information is the family page, as the name suggests it describes the Pfam family records. To navigate to the family pages the users can also enter the Pfam identifier or the accession number in the keyword search box. There are various tabs for specific information such as alignments, curation and models, domain organization or architectures, functional annotation, HMM logo, interactions, species distribution, structure and Trees (Finn et al., 2009).

Keyword Search

The search box in the page header of each page of Pfam website can be used for keyword search. This type of searching can be done by entering various types of queries. Some of the most common ways are: Gene Ontology IDs and terms, HEADER and TITLE fields from PDB entries, InterPro entry abstracts, sequence entry description and species fields in UniProt and text fields in Pfam entries (Finn et al., 2009).

Using Pfam

To determine the domain architecture of the protein of interest, the user needs to search that particular protein sequence against the Pfam library of HMMs. If the protein is not recognized by Pfam then the user shall paste the complete protein sequence in search page. The sequence will be searched against the HMMs present in the database and the matches will be displayed as the result. If there a large number of sequences to be searched then batch upload facility can be used. The user needs to upload a file containing all the sequences in the FASTA format. The results will be emailed back to the user usually 48 hours after submission. If there are a considerably very huge number of sequences then Pfam searches can be performed locally by the user through the use of the 'pfam_scan.pl' script. This technique requires the user to have additional data files from website, HMMER3 software and Pfam HMM libraries. Pfam Alyzer is a tool which can be used by the user to identify the proteins which contain a specific combination of domains and to specify particular species and the evolutionary distances allowed between domains (Finn et al., 2009).

Scores in Pfam

E-Values and Bit-Scores

HMMER3 calculates E-values (expectation values). It can be defined as a count of hits expected in order to have a score which is either equal to or better than the value by chance. E-value is considered good if it is much less than 1. Since E-values are dependent on the size of the database searched, hence a second system is used for retaining Pfam models. This in-house system is based on a bit score and hence does not depend on the size of the database searched for. Bit score gathering (GA) threshold is calculated for each Pfam family and thereby is set manually in such a manner that all sequences scoring either exact or overhead this threshold appear in full alignment.

Sequence vs. Domain Scores

HMMER3 calculates two kinds of scores known as sequence score and domain score. Sequence score provides the score for the complete sequence. Domain sequence provides the score for the domain(s) on that particular sequence. The sequence score can be defined as an aggregate score of a sequence which is aligned to HMM model. In a sequence, a single domain is said to be existing when scores are the same. Therefore, the result of these multiple illustrations of domain enhances confidence thus concluding that sequence belongs to a specific family of protein.

Table 9 provides a list of most commonly used terms in Pfam's website.

Table 9. Most commonly used terms on Pfam's website

Term	Inference
Alignment coordinates	They are those coordinates over which the HMMER3 software is sure that the alignment of profile HMM and the sequence is correct.
Architecture	It is the set of domains present on a protein.
Clan	It is a set of those Pfam records which share either profile-HMM, sequence or structure similarity.
Domain	It is a structural unit.
Domain score	It is the score of a single domain aligned to an HMM.
DUF	It is the domain of unknown function.
Envelope coordinates	They are those coordinates on which HMMER3 software has found matches to probabilistically lie.
Family	It is a set of related protein regions.
Full alignment	It is an alignment of those related sequences which have higher score as compared to that which was set manually for HMMs of a particular Pfam entry.
Gathering threshold (GA)	It is a search threshold which is used to build the full alignment.
HMMER	It is a set of programs for building as well as searching HMMs.
Hidden Markov model (HMM)	It is a probabilistic model.
IPfam	It contains information about domain-domain interactions observed in PDB entries.
Metaseq	It is a set of sequences which have been collected from different metagenomics datasets.
Motif	It is a short unit found outside globular domains.
Noise cutoff (NC)	It is the bit score of highest scoring match that was not in the full alignment.
Pfam-A	It is HMM based and was built using small number of sequences.
Posterior probability	It determines the accuracy for state prediction of profile HMM models.
Repeat	Repeat is a term used for a short unit which becomes stable when there are multiple copies but in isolation it is unstable.
Seed alignment	It refers to alignment of representative sequences set.
Sequence score	It is the combined score of a sequence aligned to a HMM.
Trusted cutoff (TC)	It is the bit score of lowest scoring match that was in the full alignment.

Finn et al., 2009.

PROSITE

Prosite is a protein pattern database which was created in 1988 by Amos Bairoch and belongs to Swiss Institute of Bioinformatics. It includes the basic patterns which are found in incomplete protein sequences, for instance, the specific functional or structural domains. Generally, patterns are found using multiple sequence alignment and then further processed according to the database (IASRI, (N.D.)). Protein motifs as well as patterns in this database are determined as regular expressions. Each record holds two forms of data that include patterns and relative descriptive text. A line commencing with "PA" declares the expression. References, as well as association for all the protein sequences that comprise of the pattern, are also revealed. Documentation files contain the descriptive text which is connected with the accession number using the expression data (IASRI, (N.D.)).

The data entries describe the protein domains, families and functional sites. In addition, they also have the information about the amino acid patterns as well as profiles. After manual curation by a team of Swiss Institute of Bioinformatics, the data is incorporated into Swiss-Prot protein annotation. It is involved in recognizing possible specific functions of newly discovered proteins and analysis of known proteins for previously undetermined action. It also suggests various implements for protein sequence analysis and detection of motif present within the protein sequence. It is part of the ExPASy proteomics analysis servers as well (De Castro et al., 2006; Hulo et al., 2008).

ProRule, is a database that is built on the domain details of PROSITE. Supplementary details for functionally or structurally critical amino acids are also provided. The ProRule comprises information of biologically meaningful residues which help in determining the function of protein. Hence automatic generation of annotation based on PROSITE motifs is possible (Sigrist et al., 2005).

KYOTO ENCYCLOPEDIA OF GENES AND GENOMES (KEGG)

KEGG can be defined as a collection of databases dealing with chemical substances, biological pathways, diseases, and drugs. It has complete genome sequences for both eukaryotes and prokaryotes (Kanehisa & Goto, 2000; Kanehisa, 1997). It is a key resource for the Japanese Genome Net service that accomplishes the challenges of defining the functional aspect relationships as well as the information related to genomes of the organism (Turenne, 2009; Rao, Das, & Umari, 2009; Morya, Dewaker, Mecarty, & Singh, 2010). KEGG is employed for analysis of data in various omic studies like genomics, meta-genomics, metabolomics, etc. (Kanehisa & Goto, 2000).

Various Databases in KEGG

KEGG can be described as a "computer representation" of a biological system (Co-oray, 2012). This concept is implemented by the databases of KEGG (Kanehisa et al., 2014). Databases can be classified on the basis of the type of information they store into Chemical, Genomic, Health and Systems.

Classification on the Basis of Chemical Data Stored

In this category, we have various sub-types which are: Compound, Enzyme, Glycan, Rclass, Reaction, and Rpair. These databases are collectively referred as KEGG LIGAND. The organization of data is done on the basis of the involved chemical networks. KEGG COMPOUND as well as KEGG GLYCAN contain data of various chemical compounds as well as glycans respectively (Hashimoto et al., 2006). KEGG REACTION contains data of multiple chemical reactions. It has two auxiliary reaction databases associated with it and these are RPAIR for reactant pair alignments and RCLASS for reaction class (Muto et al., 2013). KEGG ENZYME has information on enzyme nomenclature (Goto, Nishioka, & Kanehisa, 1999).

Classification on the Basis of Genomic Data Stored

In this category, we have three sub-types which are as Genome, Genes, and Orthology. KEGG GENOME database contains data for complete genomes of various organisms. KEGG GENE database consists of information on genes as well as proteins which are present in complete genomes. Genes are linked with various annotations which are related to pathway maps, modules, and BRITE hierarchies. Initially, pathway maps are drawn out using the experimental data obtained from specific organisms. Then, later on, they are generalized so that the maps become applicable to all organisms. This generalization is justified because there are multiple organisms which share similar pathways containing genes which are functionally identical which are known as orthologs. All genes are first assembled into orthologs and then these orthologs are stored in the KEGG ORTHOLOGY (KO) database (Kanehisa et al., 2014).

Classification on the Basis of Health Data Stored

In this category, we have four sub-types which are: Disease, Drugs, Environ and Medicus. KEGG PATHWAY database includes both the normal and abnormal states of various biological systems. Due to lack of agreement and understanding on the molecular basis of many diseases, it is not feasible to draw disease pathway maps for various diseases. KEGG DISEASE database has considered an alternative approach in which it simply registers known genetic and environmental aspects of diseases. KEGG DRUG database stores the requirements of drugs which are permitted in Europe, Japan, and the USA. The distinction between the drugs can be made on the basis of chemical components and/or structures, target molecules accompanying the drug in focus, metabolic enzymes and other network data of molecular interaction. This empowers a unified analysis of drug relations along with information of complete genome. Crude drugs and other health-related constituents are kept in KEGG ENVIRON database. KEGG MEDICUS is a database in health information category and it takes into account the package information of all advertised drugs in Japan (Kanehisa, Goto, Furumichi, Tanabe, & Hirakawa, 2010).

Classification on the Basis of Systems Data Stored

In this category we have three sub-types which are as follows:

1. **KEGG PATHWAY:** Pathway database is one of the fundamental KEGG resources which stores computerized information on the interaction of molecular networks. It is also known as the wiring diagram database. The objective was to enable the interpretation of genome sequence data through this database. It contains KEGG pathway maps which have experimental information on different pathways present in cell or organism. Through a molecular network, the genes in genomic sequence can be linked to gene products in the pathway. It also determines the pathways that are most likely to be encoded in a genome. The pathway maps are classified into the following sections: cellular processes, environmental information processing, genetic information processing, metabolism and organismal systems. Cellular processes deal with death, growth and membrane functions of the cell. Environmental information processing considers transportation through membranes and different mechanisms of signal transduction. Genetic information processing takes into account the transcription, translation, replication and repair mechanisms. Metabolism segment has visually drawn global maps of metabolic pathways. Organismal systems include various systems in an organism like endocrine, immune and nervous system.
2. **KEGG MODULE:** It highlights the functional units present in a pathway map. For instance, it will store the sub-pathways which have been reported to be conserved among specific molecular complexes as well as organisms. The gene sets present in these modules can be linked with various specific metabolic capacities as well as phenotypic features.

3. **KEGG BRITE:** It can be defined as an ontology database that contains a categorized arrangement of various biological entities. It can have many dissimilar relationships which are in contrast with KEGG PATHWAY as it contains only molecular interactions and reactions.

KEGG Search

It accepts an entry identifier in order to save the resultant entry. The identifier may be in the form of a database-dependent prefix followed by a five-digit number. If DBGET mode is implemented, then DBGET search will be made against the entire KEGG database. MEDICUS is a default mode in the KEGG MEDICUS page. A keyword search also implemented in order to get the required information (www.genome.jp/kegg/).

CONCLUSION

The book chapter provides an overview of the most common databases on the information and analysis provided by each database, information retrieval system and formats available, along with utility of the database to its users. The diversity of databases makes it challenging to identify which database should be used to solve a particular problem because database nomenclature is not standardized and data formats are also varying. Hence databases struggle with data redundancy and data inconsistencies. Different strategies have been proposed to prioritize choice of a particular database on the basis of the purpose of usage.

REFERENCES

Altairac, S. (2006). Naissance d'une banque de données: Interview du prof. Amos Bairoch. *Protéines à la Une.*

Altschul, S. F., Gish, W., Miller, W., Myers, E. W., & Lipman, D. J. (1990). Basic local alignment search tool. *Journal of Molecular Biology*, *215*(3), 403–410. doi:10.1016/S0022-2836(05)80360-2 PMID:2231712

Amid, C., Birney, E., Bower, L., Cerdeño-Tárraga, A., Cheng, Y., Cleland, I., & Hunter, C. et al. (2011). Major submissions tool developments at the European Nucleotide Archive. *Nucleic Acids Research*, gkr946. PMID:22080548

Andreeva, A., Howorth, D., Brenner, S. E., Hubbard, T. J., Chothia, C., & Murzin, A. G. (2004). SCOP database in 2004: Refinements integrate structure and sequence family data. *Nucleic Acids Research*, *32*(suppl 1), D226–D229. doi:10.1093/nar/gkh039 PMID:14681400

Apweiler, R., Bairoch, A., & Wu, C. H. (2004). Protein sequence databases. *Current Opinion in Chemical Biology*, *8*(1), 76–80. doi:10.1016/j.cbpa.2003.12.004 PMID:15036160

Apweiler, R., Bairoch, A., Wu, C. H., Barker, W. C., Boeckmann, B., Ferro, S., & Magrane, M. et al. (2004). UniProt: The universal protein knowledgebase. *Nucleic Acids Research*, *32*(suppl 1), D115–D119. doi:10.1093/nar/gkh131 PMID:14681372

Bairoch, A., & Apweiler, R. (1996). The SWISS-PROT protein sequence data bank and its new supplement TREMBL. *Nucleic Acids Research*, *24*(1), 21–25. doi:10.1093/nar/24.1.21 PMID:8594581

Bairoch, A. M. (2000). Serendipity in bioinformatics, the tribulations of a Swiss bioinformatician through exciting times! *Bioinformatics (Oxford, England)*, *16*(1), 48–64. doi:10.1093/bioinformatics/16.1.48 PMID:10812477

Barrett, T., Troup, D. B., Wilhite, S. E., Ledoux, P., Rudnev, D., Evangelista, C., & Edgar, R. et al. (2007). NCBI GEO: Mining tens of millions of expression profiles—database and tools update. *Nucleic Acids Research*, *35*(suppl 1), D760–D765. doi:10.1093/nar/gkl887 PMID:17099226

Bateman, A., Coin, L., Durbin, R., Finn, R. D., Hollich, V., Griffiths-Jones, S., & Sonnhammer, E. L. et al. (2004). The Pfam protein families database. *Nucleic Acids Research*, *32*(suppl 1), D138–D141. doi:10.1093/nar/gkh121 PMID:14681378

Berman, H., Henrick, K., & Nakamura, H. (2003). Announcing the worldwide protein data bank. *Nature Structural & Molecular Biology*, *10*(12), 980–980. doi:10.1038/nsb1203-980 PMID:14634627

Berman, H. M., Westbrook, J., Feng, Z., Gilliland, G., Bhat, T., Weissig, H., & Bourne, P. E. et al. (2000). The protein data bank. *Nucleic Acids Research*, *28*(1), 235–242. doi:10.1093/nar/28.1.235 PMID:10592235

Birney, E., & Clamp, M. (2004). Biological database design and implementation. *Briefings in Bioinformatics*, *5*(5), 31–38. doi:10.1093/bib/5.1.31 PMID:15153304

Buneman, P., Davidson, S. B., Hart, K., Overton, C., & Wong, L. (1995). *A data transformation system for biological data sources*. Academic Press.

Chandonia, J.-M., Andreeva, A., Howorth, D., Brenner, S. E., Hubbard, T. J., Chothia, C., & Murzin, A. G. (2007). *Data growth and its impact on the SCOP database: new developments*. Berkeley, CA: Ernest Orlando Lawrence Berkeley National Laboratory.

Cochrane, G., Akhtar, R., Aldebert, P., Althorpe, N., Baldwin, A., Bates, K., & Browne, P. et al. (2008). Priorities for nucleotide trace, sequence and annotation data capture at the Ensembl Trace Archive and the EMBL Nucleotide Sequence Database. *Nucleic Acids Research*, *36*(suppl 1), D5–D12. doi:10.1093/nar/gkm1018 PMID:18039715

Codd, E. F. (1970). A relational model of data for large shared data banks. *Communications of the ACM*, *13*(6), 377–387. doi:10.1145/362384.362685

Consortium, U. (2010). The universal protein resource (UniProt) in 2010. *Nucleic Acids Research*, *38*(suppl 1), D142–D148. doi:10.1093/nar/gkp846 PMID:19843607

Consortium, U. (2011). Ongoing and future developments at the Universal Protein Resource. *Nucleic Acids Research*, *39*(suppl 1), D214–D219. doi:10.1093/nar/gkq1020 PMID:21051339

Conte, L. L., Ailey, B., Hubbard, T. J., Brenner, S. E., Murzin, A. G., & Chothia, C. (2000). SCOP: A structural classification of proteins database. *Nucleic Acids Research*, *28*(1), 257–259. doi:10.1093/nar/28.1.257 PMID:10592240

Conte, L. L., Brenner, S. E., Hubbard, T. J., Chothia, C., & Murzin, A. G. (2002). SCOP database in 2002: Refinements accommodate structural genomics. *Nucleic Acids Research*, *30*(1), 264–267. doi:10.1093/nar/30.1.264 PMID:11752311

Cooray, D. M. P. N. S. (2012). Molecular biological databases: Evolutionary history, data modeling, implementation and ethical background. *Sri Lanka. Journal of Biomedical Informatics*, *3*(1), 2–11. doi:10.4038/sljbmi.v3i1.2489

Day, R., Beck, D. A., Armen, R. S., & Daggett, V. (2003). A consensus view of fold space: Combining SCOP, CATH, and the Dali Domain Dictionary. *Protein Science*, *12*(10), 2150–2160. doi:10.1110/ps.0306803 PMID:14500873

De Castro, E., Sigrist, C. J., Gattiker, A., Bulliard, V., Langendijk-Genevaux, P. S., Gasteiger, E., & Hulo, N. et al. (2006). ScanProsite: Detection of PROSITE signature matches and ProRule-associated functional and structural residues in proteins. *Nucleic Acids Research*, *34*(suppl 2), W362–W365. doi:10.1093/nar/gkl124 PMID:16845026

Finn, R. D., Marshall, M., & Bateman, A. (2005). iPfam: Visualization of protein–protein interactions in PDB at domain and amino acid resolutions. *Bioinformatics (Oxford, England)*, *21*(3), 410–412. doi:10.1093/bioinformatics/bti011 PMID:15353450

Finn, R. D., Mistry, J., Schuster-Böckler, B., Griffiths-Jones, S., Hollich, V., Lassmann, T., & Durbin, R. et al. (2006). Pfam: Clans, web tools and services. *Nucleic Acids Research*, *34*(suppl 1), D247–D251. doi:10.1093/nar/gkj149 PMID:16381856

Finn, R. D., Mistry, J., Tate, J., Coggill, P., Heger, A., Pollington, J. E., & Forslund, K. et al. (2009). The Pfam protein families database. *Nucleic Acids Research*. PMID:19920124

Garg, N., Pundhir, S., Prakash, A., & Kumar, A. (2008). PCR primer design: DREB genes. *J Comput Sci Syst Biol*, *1*, 21-40.

Goto, S., Nishioka, T., & Kanehisa, M. (1999). LIGAND database for enzymes, compounds and reactions. *Nucleic Acids Research*, *27*(1), 377–379. doi:10.1093/nar/27.1.377 PMID:9847234

Hadley, C., & Jones, D. T. (1999). A systematic comparison of protein structure classifications: SCOP, CATH and FSSP. *Structure (London, England)*, *7*(9), 1099–1112. doi:10.1016/S0969-2126(99)80177-4 PMID:10508779

Hamm, G. H., & Cameron, G. N. (1986). The EMBL data library. *Nucleic Acids Research*, *14*(1), 5–9. doi:10.1093/nar/14.1.5 PMID:3945550

Hamosh, A., Scott, A. F., Amberger, J. S., Bocchini, C. A., & McKusick, V. A. (2005). Online Mendelian Inheritance in Man (OMIM), a knowledgebase of human genes and genetic disorders. *Nucleic Acids Research*, *33*(suppl 1), D514–D517. doi:10.1093/nar/gki033 PMID:15608251

Hasegawa, H. (2008). *Genome Databases Current Implementation Practices*. Retrieved in October.

Hashimoto, K., Goto, S., Kawano, S., Aoki-Kinoshita, K. F., Ueda, N., Hamajima, M., & Kanehisa, M. et al. (2006). KEGG as a glycome informatics resource. *Glycobiology*, *16*(5), 63R–70R. doi:10.1093/glycob/cwj010 PMID:16014746

Heger, A., Wilton, C. A., Sivakumar, A., & Holm, L. (2005). ADDA: A domain database with global coverage of the protein universe. *Nucleic Acids Research*, *33*(suppl 1), D188–D191. doi:10.1093/nar/gki096 PMID:15608174

Hubbard, T. J., Ailey, B., Brenner, S. E., Murzin, A. G., & Chothia, C. (1999). SCOP: A Structural Classification of Proteins database. *Nucleic Acids Research*, *27*(1), 254–256. doi:10.1093/nar/27.1.254 PMID:9847194

Hulo, N., Bairoch, A., Bulliard, V., Cerutti, L., Cuche, B. A., De Castro, E., & Sigrist, C. J. et al. (2008). The 20 years of PROSITE. *Nucleic Acids Research*, *36*(suppl 1), D245–D249. doi:10.1093/nar/gkm977 PMID:18003654

Kanehisa, M. (1997). A database for post-genome analysis. *Trends in Genetics*, *9*(13), 375–376. doi:10.1016/S0168-9525(97)01223-7 PMID:9287494

Kanehisa, M. (2013). Chemical and genomic evolution of enzyme-catalyzed reaction networks. *FEBS Letters*, *587*(17), 2731–2737. doi:10.1016/j.febslet.2013.06.026 PMID:23816707

Kanehisa, M., & Goto, S. (2000). KEGG: Kyoto encyclopedia of genes and genomes. *Nucleic Acids Research*, *28*(1), 27–30. doi:10.1093/nar/28.1.27 PMID:10592173

Kanehisa, M., Goto, S., Furumichi, M., Tanabe, M., & Hirakawa, M. (2010). KEGG for representation and analysis of molecular networks involving diseases and drugs. *Nucleic Acids Research*, *38*(Database suppl 1), D355–D360. doi:10.1093/nar/gkp896 PMID:19880382

Kanehisa, M., Goto, S., Sato, Y., Kawashima, M., Furumichi, M., & Tanabe, M. (2014). Data, information, knowledge and principle: Back to metabolism in KEGG. *Nucleic Acids Research*, *42*(D1), D199–D205. doi:10.1093/nar/gkt1076 PMID:24214961

Kumar, S. (2005). *Bioinformatics Web*. Retrieved November 2015, from http://www.bioinformaticsweb.net/data.html

Leinonen, R., Akhtar, R., Birney, E., Bower, L., Cerdeno-Tárraga, A., Cheng, Y., & Gibson, R. et al. (2010). The European nucleotide archive. *Nucleic Acids Research*.

Lindberg, D. (2000). Internet access to the National Library of Medicine. *Effective Clinical Practice*, *3*(5), 256. PMID:11185333

Maglott, D., Ostell, J., Pruitt, K. D., & Tatusova, T. (2007). Entrez Gene: Gene-centered information at NCBI. *Nucleic Acids Research*, *35*(suppl 1), D26–D31. doi:10.1093/nar/gkl993 PMID:17148475

Mcwilliam, H., Valentin, F., Goujon, M., Li, W., Narayanasamy, M., Martin, J., & Lopez, R. et al. (2009). Web services at the european bioinformatics institute-2009. *Nucleic Acids Research*, *37*(suppl 2), W6–W10. doi:10.1093/nar/gkp302 PMID:19435877

Meyer, E. E. (1997). The first years of the Protein Data Bank. *Protein Science*, *6*(7), 1591–1597. doi:10.1002/pro.5560060724 PMID:9232661

Morya, V., Dewaker, V., Mecarty, S., & Singh, R. (2010). In silico analysis of metabolic pathways for identification of putative drug targets for Staphylococcus aureus. *J Comput Sci Syst Biol, 3*(3), 62-69.

Muto, A., Kotera, M., Tokimatsu, T., Nakagawa, Z., Goto, S., & Kanehisa, M. (2013). Modular architecture of metabolic pathways revealed by conserved sequences of reactions. *Journal of Chemical Information and Modeling, 53*(3), 613–622. doi:10.1021/ci3005379 PMID:23384306

ODonovan, C., Martin, M. J., Gattiker, A., Gasteiger, E., Bairoch, A., & Apweiler, R. (2002). High-quality protein knowledge resource: SWISS-PROT and TrEMBL. *Briefings in Bioinformatics, 3*(3), 275–284. doi:10.1093/bib/3.3.275 PMID:12230036

Orengo, C. A., Michie, A., Jones, S., Jones, D. T., Swindells, M., & Thornton, J. M. (1997). CATH–a hierarchic classification of protein domain structures. *Structure (London, England), 5*(8), 1093–1109. doi:10.1016/S0969-2126(97)00260-8 PMID:9309224

Protein_Data_Bank. (n.d.). *Protein Data Bank Wiki*. Retrieved from https://en.wikipedia.org/wiki/Protein_Data_Bank

Pruitt, K. D., Tatusova, T., & Maglott, D. R. (2005). NCBI Reference Sequence (RefSeq): A curated non-redundant sequence database of genomes, transcripts and proteins. *Nucleic Acids Research, 33*(suppl 1), D501–D504. doi:10.1093/nar/gki025 PMID:15608248

Rao, V., Das, S., & Umari, E. (2009). Glycomics Data Mining. *J Comput Sci Syst Biol, 2*, 262–265.

Sahoo, G. C., Rani, M., Dikhit, M. R., Ansari, W. A., & Das, P. (2009). Structural modeling, evolution and ligand interaction of KMP11 protein of different leishmania strains. *J Comput Sci Syst Biol, 2*, 147–158.

Schäffer, A. A., Aravind, L., Madden, T. L., Shavirin, S., Spouge, J. L., Wolf, Y. I., & Altschul, S. F. et al. (2001). Improving the accuracy of PSI-BLAST protein database searches with composition-based statistics and other refinements. *Nucleic Acids Research, 29*(14), 2994–3005. doi:10.1093/nar/29.14.2994 PMID:11452024

Schuler, G. D. (1997). Pieces of the puzzle: Expressed sequence tags and the catalog of human genes. *Journal of Molecular Medicine, 75*(10), 694–698. doi:10.1007/s001090050155 PMID:9382993

Schuler, G. D., Epstein, J. A., Ohkawa, H., & Kans, J. A. (1996). Entrez: Molecular biology database and retrieval system. *Methods in Enzymology*, *266*, 141–162. doi:10.1016/S0076-6879(96)66012-1 PMID:8743683

Sigrist, C. J., De Castro, E., Langendijk-Genevaux, P. S., Le Saux, V., Bairoch, A., & Hulo, N. (2005). ProRule: A new database containing functional and structural information on PROSITE profiles. *Bioinformatics (Oxford, England)*, *21*(21), 4060–4066. doi:10.1093/bioinformatics/bti614 PMID:16091411

Sillitoe, I., Lewis, T. E., Cuff, A., Das, S., Ashford, P., Dawson, N. L., & Lees, J. G. et al. (2015). CATH: Comprehensive structural and functional annotations for genome sequences. *Nucleic Acids Research*, *43*(D1), D376–D381. doi:10.1093/nar/gku947 PMID:25348408

Singh, S., Gupta, S., Nischal, A., Khattri, S., & Nath, R. (2010). Comparative Modeling Study of the 3-D Structure of Small Delta Antigen Protein of Hepatitis Delta Virus. *J Comput Sci Syst Biol*, *3*, 1-4.

Stoesser, G., Baker, W., van den Broek, A., Garcia-Pastor, M., Kanz, C., Kulikova, T., & Lopez, R. et al. (2003). The EMBL nucleotide sequence database: Major new developments. *Nucleic Acids Research*, *31*(1), 17–22. doi:10.1093/nar/gkg021 PMID:12519939

Subramanian, R., Muthurajan, R., & Ayyanar, M. (2008). Comparative Modeling an d Analysis of 3-D Structure of EMV2, a Late Embryogenesis Abundant Protein o f Vigna Radiata (Wilczek). *J Proteomics Bioinform*, *1*(8), 401–407. doi:10.4172/jpb.1000049

Suzek, B. E., Huang, H., McGarvey, P., Mazumder, R., & Wu, C. H. (2007). UniRef: Comprehensive and non-redundant UniProt reference clusters. *Bioinformatics (Oxford, England)*, *23*(10), 1282–1288. doi:10.1093/bioinformatics/btm098 PMID:17379688

Tatusova, T. A., Karsch-Mizrachi, I., & Ostell, J. A. (1999). Complete genomes in WWW Entrez: Data representation and analysis. *Bioinformatics (Oxford, England)*, *15*(7), 536–543. doi:10.1093/bioinformatics/15.7.536 PMID:10487861

Toomula, N. (2011). Biological databases-integration of life science data. *Journal of Computer Science & Systems Biology*.

Turenne, N. (2009). Data mining, a tool for systems biology or a systems biology tool. *J Comput Sci Syst Biol*, *2*(4), 216–218. doi:10.4172/jcsb.1000034e

Varsale, A., Wadnerkar, A., Mandage, R., & Jadhavrao, P. (2010). Cheminformatics. *J Proteomics Bioinform*, *3*, 253–259. doi:10.4172/jpb.1000148

Wheeler, D. L., Barrett, T., Benson, D. A., Bryant, S. H., Canese, K., Chetvernin, V., & Federhen, S. et al. (2007). Database resources of the national center for biotechnology information. *Nucleic Acids Research*, *35*(suppl 1), D5–D12. doi:10.1093/nar/gkl1031 PMID:17170002

Wu, C., & Nebert, D. W. (2004). Update on genome completion and annotations: Protein Information Resource. *Human Genomics*, *1*(3), 229. doi:10.1186/1479-7364-1-3-229 PMID:15588483

Yooseph, S., Sutton, G., Rusch, D. B., Halpern, A. L., Williamson, S. J., Remington, K., & Li, W. et al. (2007). The Sorcerer II Global Ocean Sampling expedition: Expanding the universe of protein families. *PLoS Biology*, *5*(3), e16. doi:10.1371/journal.pbio.0050016 PMID:17355171

KEY TERMS AND DEFINITIONS

Bioinformatics: The subjects, which deals with the analysis of biological data with computational approaches.

Biological Databases: The data generated through biological experiments organized using computational programs.

NCBI: National Center for Biological Information.

Software Tools: One of the primary activities in bioinformatics, which deals with the development of computer program for biological data analysis.

APPENDIX

Table 10. A list of resources that can be accessed and/or are available through EMBL-EBI

Services	Description
ArrayExpress	It contains information on experimentations related to gene expression.
BioModels Database	It contains catalog of computational models which are related to life sciences.
Chemical Entities of Biological Interest (ChEBI)	It can be considered as a database as well as ontology of molecular entities.
Clustal Omega	It performs multiple sequence alignment of nucleotide or protein sequences.
Clustal Phylogeny	It generates a phylogenetic tree based on the result given by ClustalW2 program.
Complex Portal	It stores manually curated biological macromolecular complexes from several model organisms
Ensembl project	It contains the genome databanks for eukaryotes. This is a joint venture with Wellcome Trust Sanger Institute.
Enzyme Portal	It provides detailed information for various enzymes.
Europe PubMed Central	It is a database that provides free access to biological research literature.
European Nucleotide Archive (ENA)	It provides information related to nucleotide sequencing.
Experimental Factor Ontology (EFO)	For various biomedical records, it provides an ontology of experimental variables.
Expression Atlas	This database stores the information regarding expression of genes based on different conditions.
FASTA	It is a protein sequence similarity search tool.
FingerPRINTScan	It determines PRINTS for a protein query sequence.
Gene ontology	It is an ontology of gene functions along with their processes.
GeneWise	It compares a protein sequence with a genomic DNA sequence.
GGSEARCH	It is used for searching sequences that are homologous to the desired query.
HMMER	It is a tool for protein homology search that uses profile hidden Markov models (HMMs).
IntEnz	It is an integrated relational enzyme database.
InterPro	It is a database of classification of proteins.
InterProScan 5	It searches sequences against InterPro's analytic protein signatures.
Kalign	It is a tool for multiple sequence alignment.
LALIGN	It is a tool to identify internal duplications.
MAFFT	It is a tool for multiple sequence alignment.
MEROPS	It is a database of proteolytic enzymes.

continued on next page

Table 10. Continued

Services	Description
MUSCLE	It is a tool for multiple sequence alignment.
MView	It is used to reformat a multiple sequence alignment or transform a sequence similarity search into a multiple sequence alignment.
NCBI BLAST	It is a tool for local similarity search.
Patent databases	It is a database of non-redundant patent sequences.
Pfam	It has been created to assign conserved protein domains and families.
Phobius	It is a tool used for the prediction of signal peptides and trans membrane topology from amino acid sequence of protein.
PICR	It provides mapping between protein identifier name spaces.
Pratt	It is a tool for discovering patterns in unaligned protein sequences.
PROSITE Scan	It is a tool for searching protein query sequence.
Protein Data Bank in Europe	It collects, distributes and organizes 3D structural data on biological macromolecular structures and their complexes.
Proteomics Identifications Database (PRIDE)	It can be defined as a repository of protein expression data which is determined with the help of mass spectrometry
UniProt (The Universal Protein Resource)	It contains the protein sequence and functional annotation data.

Retrieved from: http://www.ebi.ac.uk/services/all.

Chapter 5
Data Mining, Big Data, Data Analytics:
Big Data Analytics in Bioinformatics

Priya P. Panigrahi
Jaypee University of Information Technology, India

Tiratha Raj Singh
Jaypee University of Information Technology, India

ABSTRACT

In this digital and computing world, data formation and collection rate are growing very rapidly. With these improved proficiencies of data storage and fast computation along with the real-time distribution of data through the internet, the usual everyday ingestion of data is mounting exponentially. With the continuous advancement in data storage and accessibility of smart devices, the impact of big data will continue to develop. This chapter provides the fundamental concepts of big data, its benefits, probable pitfalls, big data analytics and its impact in Bioinformatics. With the generation of the deluge of biological data through next generation sequencing projects, there is a need to handle this data trough big data techniques. The chapter also presents a discussion of the tools for analytics, development of a novel data life cycle on big data, details of the problems and challenges connected with big data with special relevance to bioinformatics.

DOI: 10.4018/978-1-5225-1871-6.ch005

INTRODUCTION

Big data has undoubtedly gained much attention in every sector like science, IT, social media, etc., in the 21st century. Myriad technological revolutions are pouring the intensification of data and data gathering. In recent years we have observed a histrionic growth in data availability. For example; the number of web pages indexed by Google, which was nearly one million in 1999, have exceeded at 4.73 billion pages in 2015, and its enlargement is speeded up by the existence of the social networks (Che *et al.*, 2013, Grobelnik, 2012, Worldometers, 2014).

Why 'Big Data' Is Essential

- Numerous diverse big data programs launched.
- Increased use of sensors in all sectors like traffic patterns, purchasing behaviors, real-time inventory management, etc.
- Supermarkets handle approximately 1 million consumer transactions every single hour, which is imported into databases estimated to have nearly three petabytes of data.
- 72 hours of video are added to YouTube every minute.
- There are approximately 217 new mobile internet users every minute.
- Facebook handles 40 to 50 billion photographs from its user base.
- Twitter users send more than 150 million tweets per day.
- Biomedical computation decoding human Genome and personalized medicine.
- Social science revolution, etc. (Shaw, 2014).

As a result of this huge amount of data 'big data' has become a modern area of potential investment. According to The McKinsey Global Institute (2012), "Big data refers to data sets whose dimension is beyond the capability of usual database software tools to capture, store, manage and analyze" (Manyika, *et al.*, 2012). In Gartner June 2012 issue, Beyer and Laney stated: "Big data are high volume, high velocity, and high variety information resources that necessitate innovative procedures to enable heightened outcomes, insight discovery, and process optimization" (Beyer & Laney, 2012). Therefore, big data needs different methods, tools, and architectures to decipher novel problems and deep-rooted problems in an improved way. Some crucial factors for the evolution of big data are; accessibility of data, enlargement of storage capabilities, and enlargement of processing power, etc. Firms in most sectors have a minimum of 100 terabytes of stored data, and several have more than a petabyte. The size of big data is so vast and multifaceted that ordinary data handling applications are insufficient. The challenges comprise search, capture, data curation, exploration, storing, allocation, conception, and data privacy (Khan *et al.*, 2014). Figure 1 describes the basic workflow of big data architecture. Precision in big data can lead to more poised results and ultimately improved decisions that lead to better operational competence, cost reduction, and reduced risk. As we are in the information age, data are actuality generated from various sources other than people and servers, like video surveillance cameras, MRI scanners, wearable devices and sensors embedded into phones, set-top boxes, etc. Considering the annual growth of data generation, the digital world information generated annually will come to 44 zettabytes by the year 2020, which is nearly ten times the magnitude of the digital world in 2013 (Payberah, 2014)

Life Cycle of Big Data Procedure

The objective of big data process is to make operative strategic results manipulating the accessibility of big data. There are six major functional components of big data life cycle as depicted in Figure 2. The acquisition includes collection, cleaning, metadata generation, and managing provenance. Mining comprises alteration, normalization, filtering, accumulation, error management. Integration includes standardization, conflict handling, reconciliation, and mapping classification. Exploration contains analysis, machine learning, and conception. Interpretation requires an understanding of the domain, information of the background, identification of patterns of importance, and tractability of the process. The last decision requires decision-making skills and endless enhancement of the procedure (Torlone, 2015).

Often Big data is mixed up with data reuse; there is a misperception between the two terms. Data reuse is normally practiced for decisional purposes of data that

Figure 1. Big data facilitates institutions to collect, accumulate, and manipulate massive quantity of data at the right speed and right time

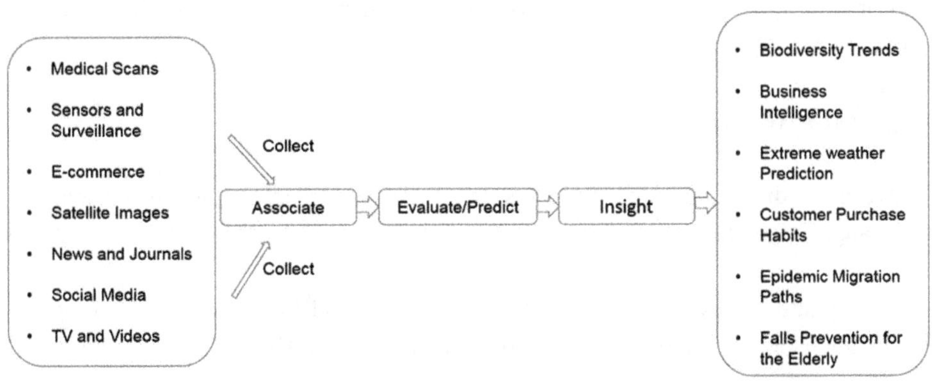

Figure 2. Functional requirements in Big data life cycle

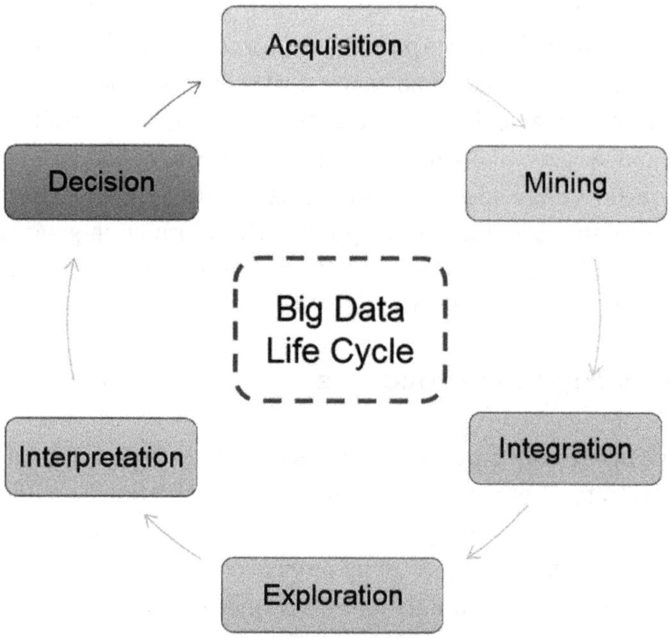

were composed consistently for transactional purposes. However, big data is associated with the size of the data assortment. Data can be reprocessed without being essentially huge, for instance, secondary usage of data from Electronic Medical Records (EMR) (Baro *et al.*, 2015). There is a major research query that can be enquired from big data sets is, whether you need to look at the full data to draw certain decisions about the properties of the data or a section of it is fair enough. The appellation big data itself holds a term correlated to mass and this in an important representative of big data. However, sampling allows the assortment of right data points inside the bigger data set to approximate the characteristics of the overall population. Big data is a revolution that will change the way we live, work, and think (Mayer-Schönberger and Kenneth, 2013).

Big Data Research Understandings

The research directed by MGI and McKinsey's Business Technology Office observed the state of big data and established the succeeding acuities (Manyika *et al.*, 2011):

1. Big data have swept into all sectors and business function and act as a significant aspect of manufacture, employment, and wealth.

2. There are five ways big data can generate importance:
    a. Big data can solve substantial assessment by making information translucent and serviceable at very high frequency.
    b. As industries generate and accumulate more transactional data in digital form, they can generate more accurate and comprehensive performance information on everything.
    c. Big data permits ever-narrower division of consumers and can consequence in much enhanced accurate personalized products or facilities.
    d. Refined analytics can noticeably improve decision-making.
    e. It can be used to advance the progress of the subsequent generation's products and facilities.
3. Eventually, Big data will become a strategic source of competition and development in individual industries.
4. It will strengthen innovative waves of efficiency growth and customer access.
5. However, the practice of big data will matter across firms, some sectors are set for better achievements.
6. There will be a deficiency of aptitude essential for industries to yield benefit from big data.
7. Numerous disputes for example; safety, privacy, intellectual property, responsibility, etc. will have to be resolved to achieve the complete benefits of big data.

BIG DATA ANALYTICS

Real-time big data architecture isn't just a procedure for gathering massive amount (petabytes or exabytes) of data in a data warehouse. It's about the potential to produce improved outcomes and take relevant actions at the correct period. It's about identifying scams like fraudulent credit card swipes or putting an advertisement on a website while someone is checking a particular web page. Big data analytics (BDA) is about linking and scrutinizing massive amount of data, thus, we can take the correct action, at an appropriate stage, and at the right time. Real-time big data analytics (RTBDA) is a route to enhance sales, increase profits and decrease marketing expenses. It indicates the beginning of a new era where technologies begin to think and respond the same as human being (Bryant *et al.*, 2008). The necessity to evaluate and influence trend data composed by industries is one of the leading drivers for BDA tools. The high-tech improvements in collecting, handling, and study of big data are as follows:

- The fast reducing price of storage and CPU power in current years.
- The flexibility and cost-effectiveness of cloud computing and data hubs for flexible computation and storing.
- The improvement in developing novel frameworks for example: Hadoop, which allows manipulators to store large quantities of data with flexible parallel processing through distributed computing systems.

The above progress has generated many differences amongst customary analytics and BDA (Watson, 2014). BDA producing a novel area of training and study named 'data science' that covers the techniques, tools, equipment, and procedures for making logical outcomes using big data. There are numerous analysis approaches, technologies, and products developed that are mostly appropriate to big data, for instance, in-database analytics, in-memory analytics, etc. Nowadays, industries usually progress analytics part from descriptive to predictive to prescriptive. The advancement is generally perceived in several BI and analytics maturity prototypes (Eckerson, 2004).

Use Cases of Big Data Analytics

Some industries have a high interest in new technologies, and thus they are quicker to assimilate RTBDA into their daily business compared to others. It's apparent that firm implementing this technology will get substantial benefits and significantly more alert and progressive in the resolutions and flexibility of their offerings. There are many industries and applications of big data use cases that are:

- Financial services providers are implementing RTBDA work frame to progress their exploration of clients and subsequently help them regulate eligibility for equity investment, secured loan, credit, insurance, etc.
- Airlines and logistic enterprises are applying RTBDA to track fuel consumption and traffic outlines across their fleets to enhance competencies and save expenses.
- Broadcasting and Telecommunication firms are using BDA to explore customer behaviors and demand outlines for an improved and effective power grid. They are using RTBDA for collecting and evaluating environmental sensor data to acquire a better understanding of infrastructure flaws and offer improved risk management intelligence.
- Healthcare personnel are also handling and sharing patient electronic health chronicles like images, treatments, statistics, etc., across various health specialists from varied sources using RTBDA. Moreover, pharmaceutical enterprises and monitoring industries are developing big data solutions to

track drug efficiency and offer more effective and smaller drug development procedures.

- Media and entertainment industries are applying RTBDA for predictive exploration of their user base and assist in making more intensive marketing and consumer analytics (Eckerson. 2011, Franks 2012, Kashyap *et al.*, 2014).

Architectures for Big Data Analytics

BDA methods have been projected with numerous architectures. The theoretical outline for a BDA project in healthcare is alike to that of a customary health analytics project. The main dissimilarity lies in which manner processing is performed. Traditional analysis of datasets can be implemented with business intelligence tool installed on an individual system, like laptop or desktop. In the case of BDA processing is shattered and performed through various nodes. Correctly achieved big data are manageable, consistent, protected, and handy. Therefore, BDA can be used in many complex scientific areas, comprising bioinformatics, biotechnology, environmental science, astronomy, medicine, genomics, biogeochemistry, etc. Availability substantial open source tools like Hadoop/MapReduce on the cloud, have stimulated the use of BDA in bioinformatics. Figure 3 specifies, the intricacy of BDA architecture. Big data in healthcare can derive from internal (clinical decision support systems, electronic health documents, etc.) as well as external sources (research laboratory, chemist's, government foundations, etc.), regularly in various formats (doc file, .csv format, Excel table, graph, image, X-ray files, relational tables, etc.) and be located in places (geographical, diverse healthcare industries etc.). The initial

Figure 3. Architectures for Big data analytics
Raghupathi and Raghupathi Health Information Science and Systems 2014, 2:3, http://www.hissjournal.com/content/2/1/3.

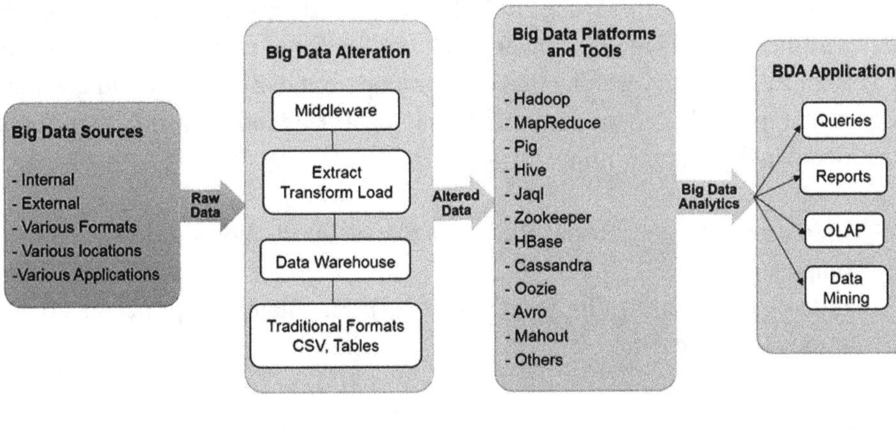

dataset is in raw format and requires to be managed or altered into a specific format, where further options are available. Middleware, one of the service-oriented method united with web applications can be used for big data alteration (Raghupathi & Kesh, 2007). Through the phases of extract, transform, and load (ETL), information from different sources is purified and organized. Based on the data types like structured or unstructured, numerous data layouts can be input to the BDA platform. In this conceptual framework, numerous assessments are made concerning the data input method, distributed design, selection of tools and analytics models (Manyika *et al.*, 2011, Khan *et al.*, 2014, Raghupathi and Raghupathi, 2014, Kashyap *et al.*, 2014).

With the advancement in computational tools, a massive amount of data can be managed without needing supercomputers and high budget. Myriad tools and techniques are obtainable for BDA, comprising Simple DB, DSMS, Google BigTable, NoSQL, Voldemort, and MemcacheDB (Chen *et al.*, 2014). Although, a lot of industries have developed distinct tools and technologies that can collect, access, and scrutinize big data in near-real time as big data varies from the customary data and cannot be kept in a distinct system. Most commonly used tools are MapReduce, Hadoop, and BigTable. These improvements have reformulated data management as they efficiently process huge volumes of data resourcefully, cost-effectively, and in a well-timed manner. In the following section, we concisely discuss data management tools and terminologies of big data.

Hadoop

Hadoop is one of the most powerful Apache project written in Java (Hadoop, 2016). '*Doug Cutting*' built-up Hadoop as an assortment of open-source projects on which the 'Google MapReduce encoding environments could be used in a scattered system. Currently, it is applied to huge quantities of data or big data. Using Hadoop, industries can able to connect data that was earlier tough to manage and scrutinize. It is used by around 63% of industries to control the massive quantity of amorphous records and events (Sys.con Media, 2011). Precisely, Hadoop can process enormously huge quantities of data with dissimilar structures format or without any structure at all. Scalability is one of essential parts of Hadoop, as it can certainly add compute nodes to the main cluster for scrutinizing big data. Hadoop project comprises of some modules those are as follows:

- **Hadoop Common:** Basic efficacies that support the additional Hadoop components.
- **Hadoop Distributed File System (HDFS):** Facilitates the principal storage for the Hadoop cluster

- **Hadoop YARN:** An outline for job scheduling and cluster resource management.
- **Hadoop MapReduce:** Based on YARN method for parallel processing of huge data sets.

Hadoop is composed of HCatalog, Pig, HBase, Zookeeper, Kafka, Oozie, and Hive. Yet, HDFS and MapReduce are the finest collective modules and renowned prototypes for Big Data. HDFS model is used when the quantity of data is excessively much for a particular machine. It is more intricate than other file structures. It splits the data into smaller chunks and allocates it through numerous servers or nodes. It stores data in blocks; the default block magnitude is 64MB. HDFS cluster comprises of two kinds of nodes; the first node is a name-node that performs as a chief node and the second one is a data node that performs as a slave node. Besides this, it also has secondary name-node. Entire HDFS records are simulated in multiples to assist the parallel processing of big data (Aho, 2012, Bakshi, 2012, Bhatnagar and Srinivasa, 2012, Katal *et al.*, 2013, Pastorelli *et al.*, 2013, Sagiroglu and Sinanc, 2013, Wang, 2011).

MapReduce

It's responsible for the interface, which is circulation of sub-tasks and the gathering of results. MapReduce supervises each and every server or node, while jobs are executing. It follows the data-parallel architecture, initially established by Google (Dean and Ghemawat, 2008). Apache Hadoop18 is a vastly used open-source application of MapReduce. Figure 4 represents the simplified architecture of MapReduce. The main node divides the data, allocates them to worker nodes, and deposits them into the global memory as key, value pairs. It functions in circles, each comprising of two stages, called map and reduce phases. A node can be used in both stages. Both phases comprise of three main states namely input, computation, and output. Amongst two succeeding phases, one synchronization barrier exists. In the course of synchronization, the local memory of one node is removed and transcribed onto the global memory. The key node can read or write onto the global memory and can also interconnect with further nodes all time. While the worker nodes can't be able to do that above during synchronization. MapReduce architecture executes well, once the data volume is vast. It offers fault-tolerance by repeating the process on another node (not on failing node). There are some limitations in MapReduce architecture that are; difficulties connecting high computational needs among data, it can't be used to direct iterative computations and becomes incompetent with high Input/Output overhead. Twister boosts the iterative computations on the architecture with the in-memory computations but, again it has fault-tolerant problems (Ekanayake

Figure 4. MapReduce architectures

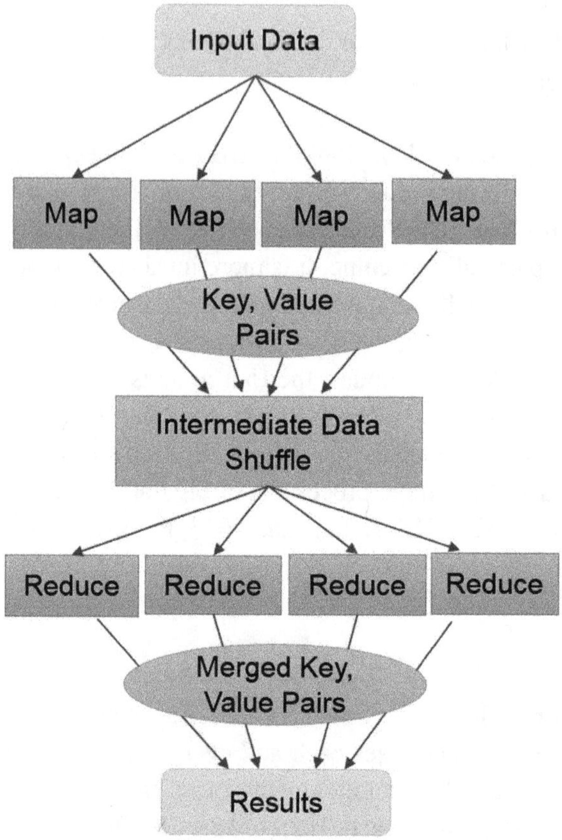

et al., 2010). Resilient Distributed Database optimizes both in-memory processing and fault-tolerance via reforming a faulty barrier if node failure takes place (Zaharia *et al.*, 2012).

- **PIG and PIG Latin:** This high-level scripting language is designed to integrate all kinds of data like structured, unstructured, etc. It consists of two main units: the programming language is named Pig Latin, and the runtime version where the Pig Latin script is executed. It is highly flexible than Hive.
- **Hive:** It is a runtime Hadoop support architecture that controls Structure Query Language (SQL). It is a separate module in the Hadoop network and develops its specific query language called as HQL or HiveQL. HQL is compiled by MapReduce and supports user-defined functions.

- **Chukwa:** It assembles and practices data from dispersed systems and retains them in Hadoop. Chukwa is an independent component, which is involved in the circulation of Apache Hadoop.
- **Jaql:** It is an efficient declarative query language intended to process huge data. Jaql transforms complex queries into simplified queries to support parallel processing, involving in MapReduce jobs.
- **HCatalog:** It keeps metadata and produces tables for big data. HCatalog is based on Hive meta store and incorporates it with other modules, for example, MapReduce and Pig. It simplifies customer statement using HDFS data. HCatalog is also a basis for data allocation amongst tools and execution platforms.
- **Zookeeper:** It permits a centralized structure having numerous facilities like enabling synchronization through a cluster of servers. BDA apply these facilities to harmonize parallel processing through big clusters. It retains, configures, and tags big data. ZooKeeper is the only one distributed service that holds *master* and *slave* nodes and preserves framework data.
- **HBase:** It is a column based DBMS that is on top of HDFS. HBase expedites the performance of actions over analogous values across big data sets. It is manageable by 'application programming interfaces' (APIs) like Java, Thrift, and representational state transfer (REST). It utilizes a non-SQL procedure.
- **Cassandra:** It is a distributed database system. Cassandra is labeled as a most powerful project designed to control big data scattered through numerous servers. It is also a NoSQL system and offers consistent facility without a specific point of failure.
- **Oozie:** It is an open source project, simplifies the workflow and synchronization among job flow. Oozie is integrated into further Apache Hadoop frameworks that are, Java MapReduce, Pig, Hive. It organizes Hadoop jobs via a directed acyclic graph (DAG).
- **Flume:** It is generally used to gather and transfer log data inside and outside of Hadoop. Flume operates two channels, called sources and sinks. Sources comprise files, system logs, and Avro, however, sinks connect to HBase and HDFS. It alters each new collection of big data before it's transferred into the sink.
- **Lucene:** It is used extensively for text analytics and has been merged into numerous open source jobs. Lucene can be used in library search and complete text indexing inside a Java application.
- **Avro:** It supports data serialization facilities. Versioning and version controller are two extra valuable amenities. Avro organizes data, clear noises, and transfer data from one platform or language to other. In this case, data are

self-descriptive and every time kept based on their individual framework. Since these abilities are mainly suitable to scripting languages like Pig.

- **Mahout:** It is an archive for data mining and machine learning. Mahout is another Apache project whose aim is to produce free modules of dispersed and accessible machine learning algorithms that support BDA. It splits into four major parts: collective cleaning, classification, clustering, and mining of recurrent parallel patterns.

BIG DATA IN BIOINFORMATICS

Biological data are being produced at a remarkable rate. The progress in proteomics, genomics, metabolomics, and other types of 'omics' technologies over the previous eras tends to the tremendous quantity of data formation related to biology, generating this new era of big data. Nowadays data is known as the fourth paradigm in information technology and science. The previous three paradigms are experimental, theoretical and computational science. The alteration in this information paradigm is being motivated by the fast development in data with advances in scientific instruments (Baro *et al.*, 2015 and Greene *at al.*, 2015). Through digitization of all processes, accessibility of high throughput devices at low costs, decreasing computing price and increasing analytics throughput with escalating big data technologies data size is expanding universally. For example, the size of a solo sequenced human genome is roughly 200 gigabytes. Biologists no longer use traditional laboratories method to discover a novel biomarker, rather they trust on massive and constantly budding genomic data made accessible by numerous research groups. Technologies for capturing biological data are getting cheaper gradually with higher efficacy, for example, computerized genome sequencers. One of the leading biological data repositories 'European Bioinformatics Institute (EBI)', had around 18 petabytes of data in 2013 which increased to 40 petabytes (proteins, genes, and small molecules) in 2014. Their total storage size is doubling every year. Other groups, like National Institute of Genetics (NIG) Japan, National Center for Biotechnology Information (NCBI), etc., are also collect and process the massive amount of biological data and make them available all over the world (Ganeshbabu, 2015 & Kashyap *et al.*, 2014).

Bioinformatics data is vastly heterogeneous in nature. There is big data like DNA, RNA, and protein sequence data, gene expression data, protein-protein interaction data (PPI), gene ontology (GO), pathway data, etc., which are used greatly in bioinformatics study. Though there are more diverse kinds of data like human disease networks and disease gene link networks (essential in diagnosis disease research) are also used in bioinformatics (Nepusz *et al.*, 2012). Sequence analysis needs advanced analytic tools with progressive computing frames, to deal with the immense quantity of sequence data. Gene expression analysis is evaluated the expression intensities of millions of genes over diverse conditions. Various resources such as Gene Expression Omnibus from NCBI, Stanford Microarray Database, and ArrayExpress from EBI, etc. and some significant PPI sources are STRING, DIP, and BioGRID which contains big data kind of information. Pathway analysis is also beneficial for understanding the molecular foundation of a disease. The famous pathway data sources are Reactome, Pathway Commons, and KEGG. The GO database offers dynamic, organized, and species-independent gene ontologies for related biological processes, cellular components, and molecular functions. It maintains certain tools, like DAG-Edit, AmiGO, and OBO-Edit (Cerami, *et al.*, 2011; Croft, *et al.*; 2010, Kanehisa & Goto; 2000, Mosquera & S´anchez-Pla, 2008; Nekrutenko& Taylor, 2012; Panigrahi & Singh, 2012; Panigrahi & Singh, 2013).

Big Data Challenges in Bioinformatics

Scientists are struggling with the big data since long, and the situation is getting intense these days. Biological data are produced by various groups, and subsequently, the similar data are available in diverse forms in the public domain. These data are huge, increasing continuously and geologically spread all over the world. However, a small quantity of these data might be accessible to all; the remaining information is not accessible (not available on the internet) because of their dimension, confidentiality, charge, and further ethical issues. Agreeing to Moore's law, i.e., '*the computing power is doubling roughly for every two years, but this rate is not matching the speed with which sequencing data is accumulating*.' The remarkable progress of biological data is presently leading the advancement of technology to acquire more improved understanding from the big data. Therefore, bioinformatics has turn into a dynamic field of research because it is proficient in mining meaningful information from big data. It needs diverse string matching tools, graph analysis techniques, and database technologies to satisfy the computational challenges in the handling of big data. It also requires more operative and effectual usage of the algorithms for reassembly of essential genome, identification of pathogens and specific genes reason for antibiotic resistance and toxins, improvement of the testing process by parallel computing software technologies and high-performance databases, etc. (Babu, 2015).

Scientists need to chomp a huge quantity of data very fast and effortlessly with the use of high-performance computers. To support this immense data parallelism and distribution it is essential to enhance:

- Data handling through thousands of servers,
- Data administration through thousands of data devices, and

- Trading with novel system data Setup.

An example of cloud-based large scale BDA is Bina box for genome analysis, which is developed recently. It functions include preprocessing on genome data and a cloud-based module to do analytics on the preprocessed data. It also moderates the genome data dimension for their effective transmission to the cloud component. This cloud-based genome analytics solution is developed by '*Bina Technologies*' which claims to advance the outcomes of genome analytics as compared to the customary methodologies (Rojahn, 2012). There are numerous big data difficulties in the field of bioinformatics, and there is a critical requisite to address these problems for example; the data that the researchers are trying to comprehend is flooded with both valuable signals and noise. Major challenges are traversal challenges for big data. Other challenging issues are heterogeneity of data, the purpose of survival for biological data and There is a need to develop new and efficient technologies and computational procedures organize, store, and analyze this deluge of diversified biological data. The major focus should be on the development of efficient storage solutions with a quick data retrieval mechanism for which techniques could be hired from data warehousing and mining applications.

Algorithms and Techniques for Big Data Analytics in Bioinformatics

A wide variability of algorithms and techniques has been built-up and improved to collect, manipulate, scrutinize, and visualize big data. These algorithms and techniques developed using numerous fields comprising computer science, statistics, information Technology, economics, applied mathematics, etc. This intends that BDA has adopted a flexible and multidisciplinary methodology. Numerous traditional techniques for data analysis are still practice on BDA. Scientists are developing novel analytical algorithms and techniques for BDA. These algorithm and techniques have direct or indirect applications in bioinformatics and related research areas. Some representative algorithms and techniques are scrutinized in this segment (Chui *et al.*, 2009, Howe, 2006, Manyika *et al.*, 2011, Sawyer & Tapia, 2005, Torres, 2014).

Data Mining Algorithms

It is a set of techniques to mine valuable information from huge, imperfect, ambiguous, and noisy datasets. These techniques comprise cluster analysis, regression analysis, association analysis, statistical learning, and machine learning, etc., using database management. Few dominant data mining techniques were recognized all through the '*IEEE International Conference on Data Mining*' (Wu *et al.*, 2008), comprising k-means, SVM, Apriori, C4.5, Cart, Naive Bayes, EM, etc. These algorithms are valuable for mining research problems in BDA. There are some classical approaches such as:

- **Neural Networks:** Computational model, based on the configuration and mechanisms of biological neural networks, which elucidate nonlinear patterns in data. Neural networks can help in pattern recognition and optimization. Some of its application implicate supervised learning, and other include unsupervised learning.
- **Association Rule Learning:** Combination of techniques to find out remarkable associations from large datasets (Agrawal *et al.*, 1966). For example; hamper market analysis, where the retailer can define which products are regularly bought together and use that fact for advertising.
- **Cluster Analysis:** A statistical process for categorizing objects that into diverse clusters, according to specific guidelines and features. It is an unsupervised learning method because training data are not used (Chen *et al.*, 2014). For instance, segmenting clients into self-similar clusters for directed advertising.

Genetic Algorithms

It is based on the procedure of natural evolution, i.e., '*survival of the fittest.*' In this procedure, significant elucidations are set as 'chromosomes' that can combine and mutate. These distinct chromosomes are nominated for survival inside a molded environment that regulates the performance of every distinct in the population. These algorithms are compatible for resolving nonlinear complications. It is also known as a category of '*evolutionary algorithm.*' Applications comprise refining job programming in industries and enhancing the performance of a stock assortment.

- **Machine Learning:** It is a domain of computer science which assist in the design and improvement of algorithms that permit computers to advance performances based on observed data. A key application of machine learning algorithm is to automatically acquire to distinguish multifaceted patterns and generate intellectual resolutions based on data. Natural language handling is an instance of machine learning.
- **Network Analysis:** A group of methods used to distinguish connections among distinct nodes in a network. In social network inquiry, links among individual characters in a community evaluated, for example exactly how information moves, or who has the best impact over whom.

- **Optimization:** A collection of numerical methods applied to remodel composite systems and procedures to enhance their performance considering some objective measures like rapidity, price, or consistency. Genetic algorithms are one sample of an optimization practice.
- **Regression Analysis:** Regression analysis used to define correlations between one variable and others. It detects dependent associations among arbitrarily hidden variables based on observation. For instance; estimating sales volumes according to the different marketplace and profitable variables.
- **Sentiment Analysis:** Involve in recognizing the feature and characteristic about which sentiment is expressed, and defining the category (positive, negative, or neutral) and its strong point. For instance, corporations applying this analysis to evaluate social media to define how diverse consumer sections and investors are responding to their products and activities.
- **Statistical Analysis:** Using statistical analysis, BDA can be inferred and described. Inferential analysis can articulate decisions concerning the data substance and random differences. However, descriptive analysis can define and encapsulate datasets. Commonly, it is used in the fields of health care and economics (84).
- **Simulation:** Sculpting the performance of multifaceted systems. Normally used for forecasting and situation design. Example; Monte Carlo simulations.
- **Visualization:** Methods applied in generating pictures, graphs, or animations to link, recognize, and enhance the outcomes of BDA.
- **Correlation Analysis:** It defines the law of associations amongst practical occurrences, comprising mutual constraint, relationship, and correlative dependence then estimates and controls dataset accordingly.

With the development of recent technologies for big data handling as part of the technical environment or as professional services, it is anticipated that BDA will evolve with the much stronger phase in the near future. The most complex and wide variety of information from biology, medicine and healthcare will provide a strong platform to help this analytics grow easily and efficiently. The information presented here is a summarized version of basic as well as an advanced form of information about Big Data and its association and application in biomedical, bioinformatics and healthcare. The field is so vast and growing with a huge pace as it's difficult to accumulate all the necessary information in a single chapter, but we tried our best to provide state-of-the-art information to the readers. Authors believe that this information will provide a basic framework for the academicians and researchers working or planning to work in the growing area.

CONCLUSION

Big data is growing and will continue to grow for its importance in various fields. There are applications, and these applications will improve the lives of communities in areas such as education, health, and employment. Ultimate challenge is how BDA continue to offer profits and opportunities to users while preserving core values and principles. Big data provides many opportunities for informatics and translational studies and will be the key for effective implementation of translational research. In this current era, it is believed and stated that "translational informatics is prepared to reform human health and healthcare using extensive measurements on organisms. Therefore data-centric methods will gain adoption to discover patterns and to mark clinically significant valuation. Cloud computing could be assets to enable translational bioinformatics research for which informatics and potential health data will be needed. Translation of medical and health data into knowledge for better healthcare will be the priority, and big data will make an important aid in it. Adopting new technologies and developing underlying infrastructure, can provide the scale, measure, and performance for the planned projections of human health care. Preparation with proper planning and exact implementation will provide the opportunity to implement translational research for the betterment of mankind and big data analytics, and bioinformatics will help in this Endeavor.

REFERENCES

Agrawal, R., Imielinski, T., & Swami, A. (1993) Mining association rules between sets of items in large databases. *SIGMOD Conference*, (pp. 207-216). doi:10.1145/170035.170072

Aho, A. V. (2012). Computation and Computational thinking. *The Computer Journal*, *55*(7), 833–835. doi:10.1093/comjnl/bxs074

Bakshi, K. (2012, March). Considerations for big data: architecture and approach. In *Proceedings of the IEEE Aerospace Conference*. doi:10.1109/AERO.2012.6187357

Baro, E. (2015). Toward a Literature-Driven Definition of Big Data in Healthcare. *BioMed Research International.*

Beyer, M., & Laney, D. (2012). *The Importance of 'Big Data': A Definition.* Gartner. Retrieved from https://www.gartner.com/doc/2057415/importance-big-data-definition

Bhatnagar, S. S. V., & Srinivasa, S. (2012). *Big Data Analytics*. Springer.

Bryant, R., Katz, R., & Lazowska, E. (2008). *Big-Data Computing: Creating revolutionary breakthroughs in commerce, science and society*. Washington, DC: Computing Community Consortium.

Cerami, E. G., Gross, B. E., Demir, E., Rodchenkov, I., Babur, O., Anwar, N., & Sander, C. et al. (2011). Pathway commons, a web resource for biological pathway data. *Nucleic Acids Research*, *39*(1), D685–D690. doi:10.1093/nar/gkq1039 PMID:21071392

Che, D., Safran, M., and Peng, Z. (2013). *From Big Data to Big Data Mining: challenges, issues, and opportunities*. Springer.

Chen, M., Mao, S., & Liu, Y. (2014). Big data: A survey. *Mobile Networks and Applications*, *19*(2), 171–209. doi:10.1007/s11036-013-0489-0

Chui, M., Miller, A., & Roberts, R. (2009). Six ways to make Web 2.0 work. *The McKinsey Quarterly*.

Croft, D. (2010). Reactome: A database of reactions, pathways and biological processes. *Nucleic Acids Research*. PMID:21067998

Dean, J., & Ghemawat, S. (2004). *MapReduce: Simplified data processing on large clusters*. Sixth Symposium on Operating System Design and Implementation, San Francisco, CA. Retrieved from labs.google.com/papers/mapreduce.html

Eckerson, W. (2004). Gauge Your Data Warehousing Maturity. *DM Review*, *11*(14), 34.

Eckerson, W. (2011). *Big Data Analytics: Profiling the Use of Analytical Platforms in User Organizations*. The Data Warehousing Institute.

Ekanayake, J. (2010). Twister: a runtime for iterative mapreduce. In *Proceedings of the 19th ACM International Symposium on High Performance Distributed Computing*, (pp. 810-818). doi:10.1145/1851476.1851593

Franks, B. (2012). *Taming the Big Data Tidal Wave*. New York: Wiley. doi:10.1002/9781119204275

Ganeshbabu, K. (2015). A Study on Role of Big Data in Bioinformatics. *International Journal of Contemporary Research in Computer Science and Technology*, *1*(7), 228.

Greene, A. C. (2015). Adapting bioinformatics curricula for big data. *Briefings in Bioinformatics*, 1–8. PMID:25829469

Grobelnik, M. (2012). *Big-Dat a Tutorial* [PowerPoint slides]. Jozef Stefan Institute. Retrieved from http://www.slideshare.net/markogrobelnik/big-data-tutorial-marko-grobelnik-25-may-2012

Hadoop, A. (2016). *Hadoop*. Retrieved from http://hadoop.apache.org/

Howe, J., (2006, June). The Rise of Crowdsourcing. *Wired, 14*(6).

Kanehisa, M., & Goto, S. (2000). KEGG: Kyoto encyclopedia of genes and genomes. *Nucleic Acids Research, 28*(1), 27–30. doi:10.1093/nar/28.1.27 PMID:10592173

Kashyap, H. (2014). Big Data Analytics in Bioinformatics: A Machine Learning Perspective. *Journal of Latex Class Files, 13*(9).

Katal, A., Wazid, M., & Goudar, R. H. (2013). Big data: Issues, challenges, tools and good practices. In *Proceedings of the 6th International Conference on Contemporary Computing* (IC3 '13). IEEE.

Khan, N. (2014). Big Data: Survey, Technologies, Opportunities, and Challenges, Hindawi Publishing Corporation. *The Scientific World Journal, 2014*, 18.

Manyika, J. (2011). *Big data: The next frontier for innovation, competition, and productivity*. McKinsey Global Institute. Retrieved from http://www.mckinsey.com/insights/mgi/research/technology_and_innovation/big_data_the_next_frontier_for_innovation

Mayer-Schönberger, V., & Cukier, K. (2013). *Big Data: A Revolution that Will Transform how We Live, Work, and Think*. New York: Houghton Mifflin Harcourt Publishing Company, Business & Economics.

Mosquera, J., & Sanchez-Pla, A. (2008). Serbgo: Searching for the best go tool. *Nucleic Acids Research, 36*(2), W368–W371. doi:10.1093/nar/gkn256 PMID:18480123

Nekrutenko, A., & Taylor, J. (2012). Next-generation sequencing data interpretation: Enhancing reproducibility and accessibility. *Nature Reviews. Genetics, 13*(9), 667–672. doi:10.1038/nrg3305 PMID:22898652

Nepusz, T., Yu, H., & Paccanaro, A. (2012). Detecting overlapping protein complexes in protein-protein interaction networks. *Nature Methods, 9*(5), 471–472. doi:10.1038/nmeth.1938 PMID:22426491

Panigrahi, P. P., & Singh, T. R. (2012). Computational analysis for functional and evolutionary aspects of BACE-1 and associated Alzheimers related proteins. *International Journal of Computational Intelligence Studies, 1*(4), 322–332. doi:10.1504/IJCISTUDIES.2012.050355

Panigrahi, P. P., & Singh, T. R. (2013). Computational analysis for Alzheimers disease associated pathways and regulatory patterns using Microarray gene expression and network data reveal association with other diseases. *Journal of Theoretical Biology*, *334*, 109–121. doi:10.1016/j.jtbi.2013.06.013 PMID:23811083

Pastorelli, M. (2013). HFSP: size-based scheduling for Hadoop. In*Proceedings of the IEEE International Congress on Big Data (BigData'13)*. IEEE.

Payberah, A. H. (2014). *Introduction to Big Data* [PowerPoint slides]. Swedish Institute of Computer Science. Retrieved from https://www.sics.se/~amir/files/download/dic/introduction.pdf

Raghupathi, W., & Kesh, S. (2007). Interoperable electronic health records design: Towards a service-oriented architecture. *e-Service Journal*, *5*(3), 39–57. doi:10.2979/ESJ.2007.5.3.39

Raghupathi, W., & Raghupathi, V. (2014). Big data analytics in healthcare: Promise and potential. *Health Information Science and Systems*, *2*(1), 3. doi:10.1186/2047-2501-2-3 PMID:25825667

Rojahn, S. Y. (2012, May). *Breaking the Genome Bottleneck*. MIT Technology.

Sagiroglu, S., & Sinanc, D. (2013, May). Big data: a review. In *Proceedings of the International Conference on Collaboration Technologies and Systems (CTS '13)*. IEEE.

Sawyer, S., & Tapia, A. (2005). The sociotechnical nature of mobile computing work: Evidence from a study of policing in the United States. *International Journal of Technology and Human Interaction*, *1*(3), 1–14. doi:10.4018/jthi.2005070101

Shaw, J. (2014, March-April). *Why "Big Data" Is a Big Deal*. Retrieved from http://harvardmagazine.com/2014/03/why-big-data-is-a-big-deal

Torlone, R. (2015). *Big data: an introduction* [PowerPoint slides]. Università Roma Tre. Retrieved from http://www.dia.uniroma3.it/~torlone/bigdata/L1-Introduzione.pdf

Torres, J. (2014). *Big Data Challenges in Bioinformatics* [PowerPoint Presentation]. Barcelona, Supercomputing Center Computer Science, Department Autonomic Systems and eBusiness Platforms.

Wang, D. (2011). An efficient cloud storage model for heterogeneous cloud infrastructures. *Procedia Engineering*, *23*, 510–515. doi:10.1016/j.proeng.2011.11.2539

Watson, H. J. (2014). Tutorial: Big Data Analytics: Concepts, Technologies, and Applications. *Communications of the Association for Information Systems*, *34*, 65.

Worldometers. (2014). *Real time world statistics*. Retrieved from http://www.worldometers.info/world-population/

Wu, X., Kumar, V., Ross Quinlan, J., Ghosh, J., Yang, Q., Motoda, H., & Steinberg, D. et al. (2008). Top 10 algorithms in data mining. *Knowledge and Information Systems*, *14*(1), 1–37. doi:10.1007/s10115-007-0114-2

Zaharia, M. (2012). Resilient distributed datasets: A fault-tolerant abstraction for in-memory cluster computing. In *Proceedings of the 9th USENIX conference on Networked Systems Design and Implementation*. USENIX Association.

KEY TERMS AND DEFINITIONS

Big Data: Big Data is a data management concept which involves the data management and analysis. Five common issues are volume, variety, velocity, value, and complexity.

Data Mining: Data processing in order to give much more information that is valuable to the end users.

Chapter 6
Principles and Analysis of Biological Networks:
Biological Pathways and Network Motifs

Manika Sehgal
Institute of Microbial Technology, India

TirathaRaj Singh
Jaypee University of Information Technology, India

ABSTRACT

The biological network complexity is growing enormously and in order to reveal confined properties of these intricate networks, detection of crucial network components may assist in gaining effortless perceptive on the underlying biological processes. Analyzing complex biological pathways for their disease association is still a drawn-out process and requires an integrative approach for comprehensive examination of proteins and interactions to identify candidate markers underlying major malignancies and genetic disorders. There is a need for an amalgamated approach to annotate all the sub-components and their associated interactions in a biological system. It is anticipated that analysis of biological pathways would serve as a valuable accompaniment for analyzing biomarkers in disease pathways and will also contribute scientific knowledge towards their better understanding.

DOI: 10.4018/978-1-5225-1871-6.ch006

INTRODUCTION

In present era, the advent of human genome sequence, high-throughput techniques and massive generated information has revolutionized the biomedical research (Bentley D. R., 2006). This huge amalgamation of genomic data is rising exponentially due to employment of next generation technologies and application of modern biology (Bentley D. R. 2006; Voelkerding. K. V et al., 2009; Mamanova L, et al. 2010). However, less emphasis is remunerated on the obligatory genomic analysis; subsequently there is a need to apply extensive computational approaches that could deal abreast with the remarkable growth in generated biological data (Barabasi A. L. & Oltvai Z.N. 2004; Jeong H., et al. 2000, Jeong H. et al. 2001). Since, the human system is very complex and the ultimate challenge is to decipher numerous multifaceted processes involved in its regulation therefore biological network analysis has major relevance which assists in wide-ranging understanding of an entire system as a whole (Jeong H., et al. 2001). These biological networks present several processes such as cellular mechanisms, signaling cascades, transcriptional regulation, expression profiles and protein-protein interactions (Barabasi A. L. & Oltvai Z. N. 2004; Jeong H. et al., 2000). In Figure 1, these networks comprise of nodes as entities and edges as relationships among them. In general terms a network may be any social interaction, electric circuits, internet and food chain, etc. In biological context, these nodes may vary from a range of macromolecules for instance DNA, RNA, proteins and metabolites. The applications of networks can be observed in ecological networks, expression networks, gene regulatory networks, metabolic networks, protein interaction networks.

Figure 1. Representation of a sub-network comprising of nodes as entities and edges as relationship/association with nodes

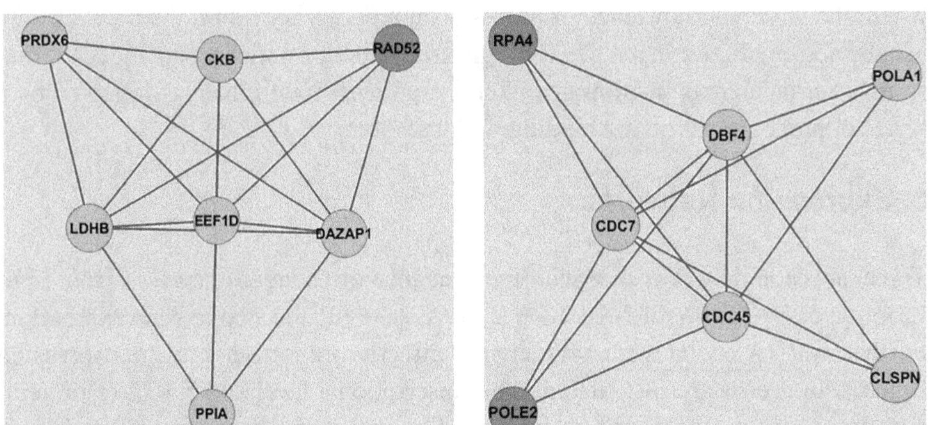

Recently, pathway analysis and reconstruction has also gained a lot of attention as it assists in unravelling unknown parameters concerning many precarious diseases. Therefore, the study of networks, their modelling, visualization and simulation are an important aspect for extensively understanding the underlying processes to make sense of the complex available data. Biological networks can be broadly classified into following types:

1. Protein-Protein Interaction (PPI) Networks

The PPI networks form the basis for a vast majority of cellular events, including signal transduction and transcriptional regulation. An extensive understanding of a biological system is feasible only once all the interactions are characterized. These interactions are very well deduced from numerous high-throughput experiments such as protein arrays, co-immunoprecipitation, microarrays and predicted from computational resources developed over time. These *in silico* tools and databases include Search Tool for the Retrieval of Interacting Genes/proteins (STRING), Biomolecular Interaction Network Database (BIND), Molecular INTeraction database (MINT) and Database of Interacting Proteins (DIP). These PPI databases help us effortlessly access the existing information for biological interactions among a variety of species. The published literature and methods used to exemplify the interaction studies are also described. These resources boost up the understanding of underlying processes which aid the scientific community for designing their future experiments.

2. Signal Transduction and Gene Regulatory Networks

The gene regulatory networks consist of clusters of genes (or gene products) with evidence of co-expression and are generally associated with cascade of processes. In this form of networks, connections usually represent degrees of co-expression where the in-depth knowledge of process is not necessary. Here, the networks are usually non-predictive in nature. The signal transduction networks are induced from some external signals (neurotransmitters, peptides, growth factors) followed by a series of processes involving binding with receptors.

3. Expression Networks

The expression networks deal with the sequence of changes occurring from transcription of DNA into mRNA which is the most regulated step in gene expression. Further, mRNA encodes for proteins i.e. directly proportional to the expression status. Most genes are regulated at the transcriptional level and 5-10% of protein-encoding genes encode regulatory proteins. The interactions between regulated genes and regulatory proteins constitute a complex web called transcriptional network or expression network.

4. Metabolic Networks

The metabolic networks comprises of all the interactions that sums up for the entire chemical reactions taking place within a cell. These processes are implicated mainly in providing energy for the vital cellular and metabolic processes. The metabolic network represents a series of chained reactions where the connections depict quantifiable contacts between the molecules. Since, a metabolic network illustrates a chained reaction, the enzymatic process is evidently elucidated where the changes are predictable downstream.

Currently, the spotlight of major resources is on the metabolic pathways, signaling and regulatory networks to fill in the research gaps and missing links between pathways. Ultimately, all these networks comprising of signal transduction, gene regulatory, metabolic and protein-protein interactions further interact with each other and build a complicated system where these networks are species–specific and may vary among different taxas (Mamanova L, et al. 2010, Jeong H., et al. 2001). There are a variety of macromolecules like enzymes, transcriptional regulators, ribosomes, other biological entities that form complex biological networks to carry out fundamental cellular functions.

BACKGROUND

For a comprehensive perspective on biological complexity and underlying intricate networks, exclusive monitoring of whole system to component level entities is required. Initially, entire pathway is analyzed which is a sequence of interactions between biochemical compounds for controlling the flow of information and energy in the cell. There are myriad of molecular compounds, molecules and cofactors that are component of a network. These include proteins, their complexes, amino acids, peptides, sugars, lipids, cofactors, prosthetic groups, metabolites and other essential compounds like Sulphur, phosphorous, iron and radicals. The analysis of biochemical pathways have become easier with the advancements in experimental technologies like microarrays, proteome analysis, yeast two-hybrid system and phosphorylations of proteins which deal with understanding of interactions among entities. Once, the PPI interactions are clearly elucidated, the reconstruction of a biological pathway becomes simpler which may further be subjected to simulation for estimating the system dynamics. This domain of pathway reconstruction and classification has led to an emerging field of network analysis that assists in wide-ranging analysis of biological pathways. This area basically employs bioinformatics approaches and utilizes computational tools for carrying out functional enrichment, annotation and defining ontologies for a biological system. The results from pathway/network analysis can also be subjected to gene ontology (GO) analysis for governing their involvement in biological processes, cellular compartments and molecular functions. The key feature for network analysis is to postulate vital conserved patterns that may be functionally significant which in turn reveals candidate genes/proteins.

The complexity of the genome becomes multi-folded as the pathways consecutively chain together to form larger pathways and intricate networks. Specific conserved patterns in these networks contribute to the disease etiology in several species. Although, cancers and multi-system defects are a major threat to mankind but still dealing with harmful consequences of these damages is not adequately feasible and also there is no absolute therapy for cancer. The understanding of major cellular processes and its implications through network analyses could help reveal important aspects regarding the same. There is an imperative need to understand these biological mechanisms and triggered signalling cascades so that damages could be sensed well in advance. Upon sensing the damage, factors involved in the regulation of pathways should be clearly elucidated along with their biological roles and positions where they exert their functions (Jeong H., et al. 2001, Jeong H., et al. 2000). Thus, to unravel the complexity hidden in intricate biological phenomenon, a pathway-level analysis for biological networks is essential for the interaction of genes/proteins/networks for understanding myriad of cellular processes.

The knowledge pertaining to interacting partners assists in examining the impact of individual proteins in the overall pathway. In general, a network has a highly complex structure comprising of thousands of nodes with connected edges representing general properties. In order to estimate these properties, absolute acquaintance on kinetics and functional dynamics is obligatory. Once, the node, edges and their properties are calibrated the ultimate challenge is to bridge both the levels. Although, the redundancy and conserved patterns can easily be exemplified further leading to identification of candidate genes/proteins in disease specific pathways. Additionally, analyzing the interaction architecture even helps in setting up a framework for mathematic models where analyzing individual components may not provide meaningful insights; therefore whole pathway analysis is the primary requisite for the study of complex systems.

MAIN FOCUS OF THE CHAPTER

In recent times, various bioinformatics and statistical approaches have been applied to thoroughly understand the structural, functional and evolutionary evidences latent in biological networks of interest. Numerous approaches are available to elucidate important findings from these intricate biological pathways, their comparisons, reconstructions and designing. Figure 2 represents the methodology that could be applied for identifying key players from a biological network. Additionally, vital network components referred as network motifs (Alon U, 2007), found in elevated frequencies that could generally be expected by chance in a pathway could also be identified, as in Figure 3. These network motifs provide statistically overrepresented sub-structures (sub-graphs) in a network and are recognized as simple building blocks of a complicated network. These network motifs play a central role in recognition and analysis of specific patterns in biological networks and yield significant insights into understanding complex biological processes involved in intricate human diseases (Kim W, et al. 2011). Major key candidates and network components or motifs can also be deduced from these complex interactions therefore confining the search to a few important genes.

To thoroughly understand the dynamics of a biological system, a comprehensive perspective on the complex interactions and its topology is required. The chapter highlights details on network properties (Assenov Y, et al. 2008) such as shortest path length, clustering coefficient, neighbourhood connectivity and other crucial parameters as depicted in Figure 4. The shortest path length $L(n,m)$ is the length of shortest path between two nodes n and m. The shortest path length distribution gives the number of node pairs (n,m) with $L(n,m) = k$ for $k = 1,2,....,z$. Whereas,

Figure 2. Methodology for detecting key candidates from the pathway

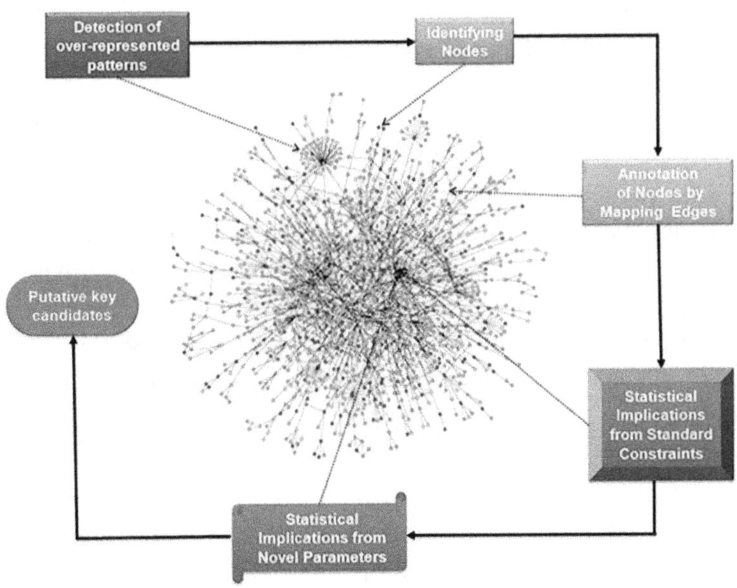

Figure 3. Identified and annotated network motifs in colorectal cancer
Sehgal M., et al., 2015.

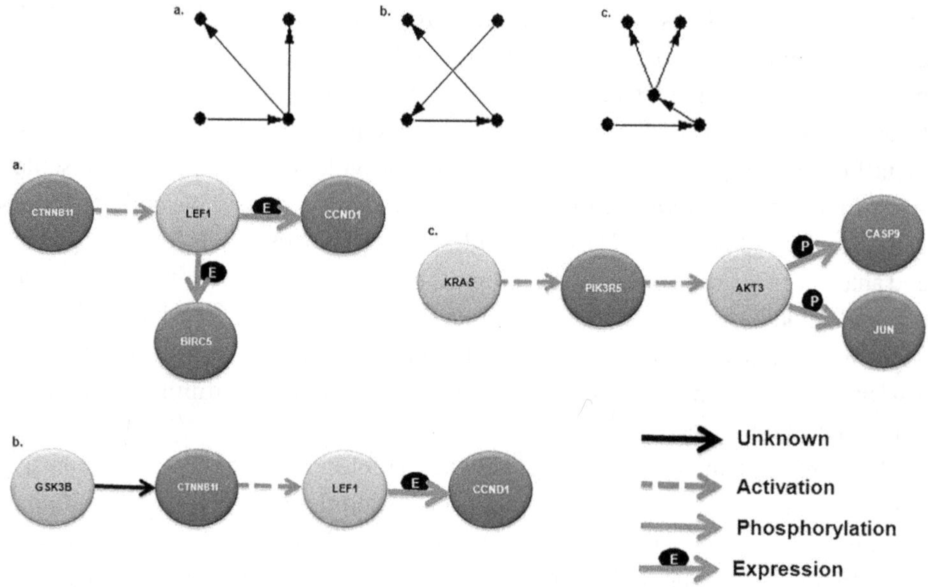

the neighborhood connectivity of a node *n* is defined as the average connectivity of all its neighbours. The neighborhood connectivity distribution gives the average of the neighborhood connectivity's of all nodes *n* with *k* neighbors for $k = 0,1,\ldots,z$. The average clustering coefficient distribution gives the average of clustering coefficients for all nodes *n* with *k* neighbors for $k = 2,\ldots,z$. In directed network, the clustering coefficient C_n of a node *n* is defined as:

$$C_n = e_n/(k_n(k_n-1))$$

where k_n is the number of neighbors of *n* and e_n is the number of connected pairs between all neighbors of *n*. The clustering coefficient is a ratio *N* / *M*, where *N* is the number of edges between the neighbors of *n*, and *M* is the maximum number of edges that could possibly exist between the neighbors of *n*. The clustering coefficient of a node always ranges from 0-1. The average clustering coefficient has already presented its utilization in metabolic networks where the modular organization has been studied.

Figure 4. Statistical parameters elucidating significance of the biological pathway

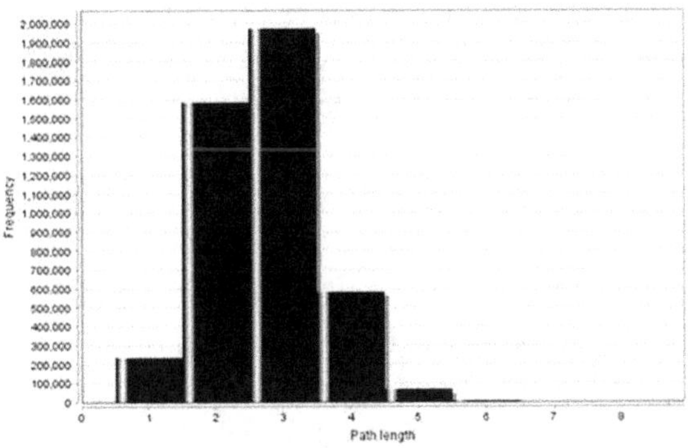

a. Shortest path length

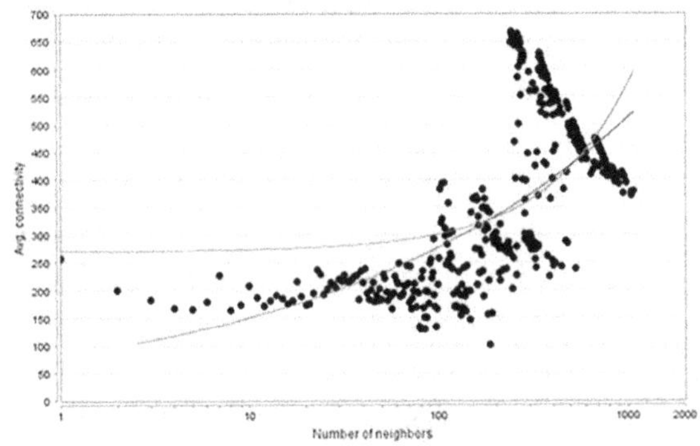

b. Neighborhood connectivity (In and Out)

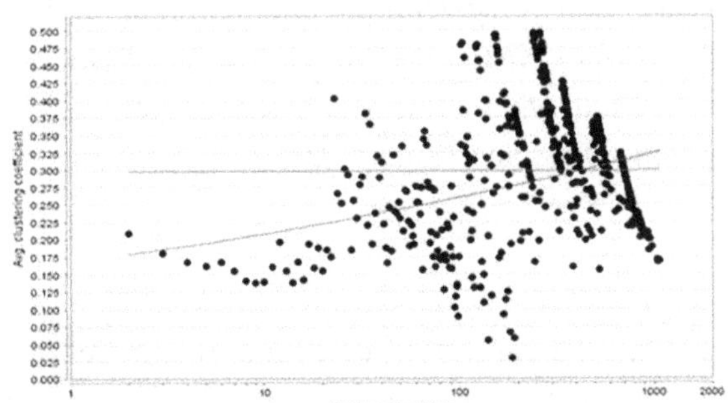

c. Average clustering coefficient

COMPUTATIONAL METHODS AND APPROACHES AVAILABLE FOR PATHWAY ANALYSIS

The rapid growth of knowledge on functional analysis of genes, proteins and cellular processes demands wide-ranging databases and novel computational methods and approaches to be devised for efficient representation of proteins as well as their functional interactions whereas the chapter extensively focus on the important available approaches. The methods and resources (Smoot M. E.,et al. 2011, Wernicke S. & Rasche F. 2006, Schreiber F. & Schwobbermeyer H. 2005, Kashtan N.,et al., 2004) for studying biological networks include:

Cytoscape

Cytoscape is a very widely used open source software platform for visualization of molecular interaction networks and biological pathways. It allows integration of the networks with single nucleotide polymorphism (SNP) data, phenotypes, gene expression profiles and other molecular states leading to annotations and functional enrichment. Cytoscape has an additional feature for loading nodes, edges and networks where different attributes could be mapped for extensive understanding of biological processes. User can easily access and analyze the expression results and GO annotations in the platform. There are various apps and plugins available for annotation of these networks incorporating data from a variety of datasets from microarrays and high-throughput studies. These plugins are generally freely available and stimulate easy visualization of complex biological data. A classical example of statistical inferences made from network analyzer plugin has already been observed in Figure 4. It has numerous features available for data integration, analysis, visualization, molecular profiling analyses, new layouts, additional file formats support and connection with databases. Cytoscape supports many standard network and annotation file formats such as SIF (Simple Interaction Format), BioPAX, PSI-MI, GraphML, KGML (KEGG XML), SBML and Gene Association. Cytoscape is considered most influential when utilized in juxtaposition with larger databases of protein-protein, protein-DNA and genetic interactions.

Fanmod

Fanmod is a tool for fast network motif detection. It is better than its counterparts (explained below) in terms of efficiency and speed when dealing with larger motifs generally more than 8-node sub- graphs. It randomly detects over-represented sub-graphs in a network called as network motifs which are recurring patterns of interactions and are statistically significant. It exploits the most efficient RAND-ESU algorithm for the detection of motifs. It provides with motifs ranging from 3-8 node sub-graphs with colored edges and have filtering done on the basis of z-score and p-values. Fanmod can be easily downloaded on a system and be used as a standalone software which performs calculations at a faster rate.

MAVisto

Like Fanmod, MAVisto is used for the exploration of motifs in a network. It provides a flexible motif search algorithm and different views for the analysis and visualization of network motifs. The MAVisto tool is written in Java and has an editor for graphs visualization and a toolkit for implementing graph algorithms. It is also available freely for academic purposes at http://mavisto.ipk-gatersleben.de/mavisto.jnlp. Usually, three different types of analyses can be performed with MAVisto by considering only edge labels, only vertex labels or both, edge and vertex labels.

Mfinder

The motif finder (mfinder) is another software tool for detecting network motifs. It identifies the network motifs in biological pathways by using two algorithms i.e. firstly it takes into account the full enumeration of sub-graphs and then sampling of sub-graphs is done for estimating the sub-graph concentrations. It employs switching method for the generation of random motifs. The standalone version can be easily downloaded from the following URL

https://www.weizmann.ac.il/mcb/UriAlon/download/network-motif-software.

The basic steps for the identification of network motifs from any of the above mentioned tool comprise of the following basic steps:

1. Identify all sub-graphs of n nodes in the network.
2. Randomize the network, while keeping the number of nodes, edges, and degree distribution unchanged.
3. Identify all sub-graphs of n nodes in the randomized version.
4. Sub-graphs that occur significantly more frequently in the real network, as compared to the randomized one, are designated as motifs.

RESOURCES AND DATABASES FOR BIOLOGICAL NETWORKS

The genome sequencing and the latest research strategies have generated huge and complex genomic and proteomic data which is available in repertoire of online resources and databases. Enormous research tools are also available for the extensive analysis of these complex interactomes. There are a variety of resources on biological networks and their analysis available online which can be classified on different means as follows:

General Pathway Repositories

This module supplies information on available general biological pathway databases. The pathway databases mainly highlight entire cascade of processes instead of individual components. Although it do provide insights into smaller components or interacting proteins but the ultimate aim is to understand the system as a whole.

BioCyc/MetaCyc

It stands for biological encyclopedia/ metabolic encyclopedia. Biocyc is a pathway/ genome database for organisms with completely sequenced genomes where species-specific pathways are inferred form MetaCyc whereas MetaCyc is a non-redundant reference database for metabolic pathways, reactions, enzymes and compounds.

Kyoto Encyclopedia of Genes and Genomes (KEGG)

KEGG is mainly considered as a common platform for consolidated set of databases that cover genomics (GENE), chemical compounds (LIGAND) and reaction networks (PATHWAY). It has a broad coverage related to other databases concerning on metabolics, signal transduction, disease, *etc*. It contains pathways from numerous species and allows only static visualization services where the pathways can be downloaded in KGML format for further annotation and analysis (Kanehisa, M.,et al. 2016).

BioCarta

It is a corporate-owned and publicly curated pathway database which provides an interactive user interface with cartoon pathway maps. BioCarta focuses mainly on human and mouse related biological pathways.

BioModels Database

BioModels is a database on published quantitative models for biochemical processes. It contains information on all models/pathways that are curated manually. It allows the models to be created in SBML format for quantitative modeling.

Some other pathway databases include PathDB, Signaling PAthway Database (SPAD) and Cytokine Signaling Pathway Database.

Protein Interaction Databases

The protein interaction databases section provides an exclusive compendium for various species-specific, function-specific, interaction type-specific and general protein-protein interaction databases.

1. **Species-Specific Databases:**
   a. **FlyNets:** Gene networks in the fruit fly,
   b. **MIPS:** Yeast Genome Database,
   c. **RegulonDB:** A DataBase On Transcriptional Regulation in E. Coli, and
   d. **PIMdb:** Drosophila Protein Interaction Map database.
2. **Function-Specific Databases:**
   a. Biocatalysis/Biodegradation Database,
   b. **BRITE:** Biomolecular Relations in Information Transmission and Expression,
   c. **COPE:** Cytokines Online Pathfinder Encyclopaedia,
   d. Dynamic Signaling Maps,
   e. **EMP:** The Enzymology Database,
   f. **FIMM:** A Database of Functional Molecular Immunology, and
   g. **CSNDB:** Cell Signaling Networks Database.
3. **Interaction Type-Specific Databases:**
   a. **DIP:** Database of Interacting Proteins,
   b. **DPInteract:** DNA-protein interactions,
   c. **Inter-Chain Beta-Sheets (ICBS):** A database of protein-protein interactions mediated by interchain beta-sheet formation,
   d. **Interact:** A Protein-Protein Interaction database, and
   e. GeneNet (Gene networks).
4. **General Protein-Protein Interaction Databases:**
   a. **BIND:** Biomolecular Interaction Network Database,
   b. **BindingDB:** The Binding Database,
   c. **MINT:** A database of Molecular INTeractions,
   d. **PATIKA:** Pathway Analysis Tool for Integration and Knowledge Acquisition,
   e. **PFBP:** Protein Function and Biochemical Pathways Project, and
   f. PIM (Protein Interaction Map) .

PATHWAY EXCHANGE FORMATS

There are numerous pathway exchange formats developed for easy exchange of information from one platform to another. The exchange formats help maintain a standard for representing pathway data or easy understanding among different researchers. Some of the exchange formats have been described below:

Extensible Markup Language (XML)

XML is a standard of representing information in a machine-readable manner. It is a division of SGML and is much better than hyper text markup language (HTML) in terms of its extensibility in defining tags that enclose data. For instance:

```
<XMLInfo>
        <someTag>Some data here</someTag>
        <anotherTag>Additional data </anotherTag>
        <attributeTag data="embedded in tag" />
</XMLInfo>
```

Systems Biology Markup Language (SBML)

SBML is a representation format based on XML for exchange and storage of computational models of biological processes. It is a machine-readable exchange format and its strength lies in the representation phenomena at the scale of biochemical reactions. Several tools use this format for standard input and output operations in order to remove chances for errors in translation and assuring a common starting point for analyses and simulations *(Keating, S. M, et al. 2006)*. SBML's framework is suitable for representing models commonly found in research on topics like gene regulation, cell signaling pathways, metabolic pathways, biochemical reactions, and many more. Example of its format is given below:

```
<?xml version="1.0" encoding="UTF-8"?>
<sbml xmlns=http://www.sbml.org/sbml/level1
level="1" version="2">
<model name="gene_network_model">
  <listOfUnitDefinitions>
        . . .
  </listOfUnitDefinitions>
<listOfCompartments>
```

```
        . . .
</listOfCompartments>
<listOfSpecies>
        . . .
</listOfSpecies>
<listOfParameters>
        . . .
</listOfParameters>
<listOfRules>
        . . .
</listOfRules>
<listOfReactions>
        . . .
</listOfReactions>
   </model>
</sbml>
```

ISSUES, CONTROVERSIES, PROBLEMS

The pathway analysis as a whole is a cumbersome task and there has been many controversies on its analysis as some researchers suggest studying individual interactions can be more beneficial instead being spellbound by the complicated system whereas others support the idea of systems biology as the biological systems are basically dynamic in nature. In the knowledge of intricate pathways/networks, regulatory DNA motifs are considered as bar codes to retrieve the building blocks which may help in gaining better insights in pathway studies. A combination of high-throughput experiments, prior knowledge and bayesian network inference is also indispensable for thorough understanding of a biological problem. Furthermore, efficient analyses on biological networks require robust computational infrastructure for specialized housing, accessing and analyses of ever-growing and amplifying complex biological data. Usually, the complete network of integrated biological pathways is the blueprint of life whereas the genome is only an archiving system of building blocks. Therefore, it is proposed that a better idea is to follow an integrative approach where both the component-levels and system-levels are taken into consideration. In order to achieve this goal, top-down and bottom-up techniques could be applied in an amalgamation.

SOLUTIONS AND RECOMMENDATIONS

The biological pathway/network analysis is a hot spot in current research domain and bioinformatics. This area has a variety of sub-fields that could be taken up by young minds to solve the mysteries of various biological problems. The fundamental themes that could be extensively deliberated include development of new algorithms, tools and comprehensive databases for understanding and visualizing pathways with noteworthy statistical inferences. The areas of research focus on valid implications of metabolic pathways from genomes, designing and development of schemas for constructing pathway related databases and classification systems for biological pathway data.

FUTURE RESEARCH DIRECTIONS

Since, major tools and resources dealing with pathway data don't have a common format for exchange of information and results, work primarily on developing exchange formats can be another valuable alternative in this area. Although systems biology markup language (SBML) solves the problem to some extent but for a consistent and straightforward perceptive, other easily understood formats are the need of the hour for exchange of pathway data. Reconstruction of biological pathways via layout algorithms from the knowledge obtained from interaction studies, simulation and molecular dynamics analysis could also yield significant insights in major biological problems. The study of protein-protein interactions has provided important insights into the functions of many of the known oncogenes, tumor suppressors, and DNA repair proteins assisting in cancer research. Moreover, pathway analysis has evidently proved its implication in the domain of pharmacogenetics research targeting specific drug transporters, drug receptors and drug targets. For any research area such as vaccine development, treatments, designing disease strategies, preventing symptoms, clinical trials and drug development to be promising, the major challenge lies in the understanding of disease mechanisms which is feasible only through widespread pathway and system-level understanding.

CONCLUSION

The chapter endows with crucial insights on major biological networks, their reconstruction and visualization, computational methods and approaches available for their analyses, online resources and databases containing information on these networks and pathways. The main challenge till date is to find appropriate targets in a diseased pathway leading to drug discovery process which has been a time consuming and expensive affair. Thus, knowledge of biological pathways will assist the readers and scientific community in gaining insights on intricate networks and available tools and resources for their study. The pathway level analysis of these networks may provide vital clues about putative but predictive biomarkers for numerous diseases. Additionally, top-down and bottom-up approaches could be applied for the annotation of complex biological networks and this functional enrichment would generate biological meaningful information. It is expected that once the biological networks are broadly analyzed, it will also provide innovative paradigms for genetic susceptibility, prevention, diagnosis and rational therapy.

REFERENCES

Alon, U. (2007). Network motifs: Theory and experimental approaches. *Nature Reviews. Genetics*, *8*(6), 450–461. doi:10.1038/nrg2102 PMID:17510665

Assenov, Y., Ramirez, F., Schelhorn, S.-E., Lengauer, T., & Albrecht, M. (2008). Computing topological parameters of biological networks. *Bioinformatics (Oxford, England)*, *24*(2), 282–284. doi:10.1093/bioinformatics/btm554 PMID:18006545

Barabasi, A. L., & Oltvai, Z. N. (2004). Network biology: Understanding the cells functional organization. *Nature Reviews. Genetics*, *5*(2), 101–113. doi:10.1038/nrg1272 PMID:14735121

Bentley, D. R. (2006). Whole-genome re-sequencing. *Current Opinion in Genetics & Development*, *16*(6), 545–552. doi:10.1016/j.gde.2006.10.009 PMID:17055251

Caspi, R., Altman, T., Billington, R., Dreher, K., Foerster, H., Fulcher, C. A., & Karp, P. D. et al. (2014). The MetaCyc Database of metabolic pathways and enzymes and the BioCyc collection of Pathway/Genome Databases. *Nucleic Acids Research*, *42*(1), D459–D471. doi:10.1093/nar/gkt1103 PMID:24225315

Jeong, H. et al.. (2000). The large-scale organization of metabolic networks. *Nature*, *407*(6804), 651–654. doi:10.1038/35036627 PMID:11034217

Jeong, H., Mason, S. P., Barabási, A.-L., & Oltvai, Z. N. (2001). Lethality and centrality in protein networks. *Nature*, *411*(6833), 41–42. doi:10.1038/35075138 PMID:11333967

Kanehisa, M., Sato, Y., Kawashima, M., Furumichi, M., & Tanabe, M. (2016). KEGG as a reference resource for gene and protein annotation. *Nucleic Acids Research, 44*(D1), D457 D462. doi:10.1093/nar/gkv1070 PMID:26476454

Kashtan, N., Itzkovitz, S., Milo, R., & Alon, U. (2004). Efficient sampling algorithm for estimating subgraph concentrations and detecting network motifs. *Bioinformatics (Oxford, England), 20*(11), 1746–1758. doi:10.1093/bioinformatics/bth163 PMID:15001476

Keating, S. M., Bornstein, B. J., Finney, A., & Hucka, M. (2006). SBMLToolbox: An SBML toolbox for MATLAB users. *Bioinformatics (Oxford, England), 22*(10), 1275–1277. doi:10.1093/bioinformatics/btl111 PMID:16574696

Kim, W., Li, M., Wang, J., & Pan, Y. (2011). Biological network motif detection and evaluation. *BMC Systems Biology, 5*(Suppl 3), S5. doi:10.1186/1752-0509-5-S3-S5 PMID:22784624

Mamanova, L., Coffey, A. J., Scott, C. E., Kozarewa, I., Turner, E. H., Kumar, A., & Turner, D. J. et al. (2010). Target-enrichment strategies for next- generation sequencing. *Nature Methods, 7*(2), 111–118. doi:10.1038/nmeth.1419 PMID:20111037

Schreiber, F., & Schwobbermeyer, H. (2005). MAVisto: A tool for the exploration of network motifs. *Bioinformatics (Oxford, England), 21*(17), 3572–3574. doi:10.1093/bioinformatics/bti556 PMID:16020473

Sehgal, M., Gupta, R., Moussa, A., & Singh, T. R. (2015). An Integrative Approach for Mapping Differentially Expressed Genes and Network Components Using Novel Parameters to Elucidate Key Regulatory Genes in Colorectal Cancer. *PLoS ONE, 10*(7), e0133901. doi:10.1371/journal.pone.0133901 PMID:26222778

Smoot, M. E., Ono, K., Ruscheinski, J., Wang, P. L., & Ideker, T. (2011). Cytoscape 2.8: New features for data integration and network visualization. *Bioinformatics (Oxford, England), 27*(3), 431–432. doi:10.1093/bioinformatics/btq675 PMID:21149340

Voelkerding, K. V., Dames, S. A., & Durtschi, J. D. (2009). Next-generation sequencing: From basic research to diagnostics. *Clinical Chemistry, 55*(4), 641–658. doi:10.1373/clinchem.2008.112789 PMID:19246620

Wernicke, S., & Rasche, F. (2006). FANMOD: A tool for fast network motif detection. *Bioinformatics (Oxford, England), 22*(9), 1152–1153. doi:10.1093/bioinformatics/btl038 PMID:16455747

KEY TERMS AND DEFINITIONS

Annotation: Annotation is defined as a measure of estimating the properties of an entity (gene, protein, metabolite, pathways) such as its functions, localizations and implicated processes. Annotation can be done through manual curation or automation depending on the availability of resources.

Network Motifs: Network motifs are defined as over-represented sub-graphs in a network. These are recurrent patterns occurring in a network with higher frequency compared to other randomly generated networks.

Chapter 7
An Overview of Biological Data Mining

Seetharaman Balaji
Manipal University, India

ABSTRACT

The largest digital repository of information, the World Wide Web keeps growing exponentially and calls for data mining services to provide tailored web experiences. This chapter discusses the overview of information retrieval, knowledge discovery and data mining. It reviews the different stages of data mining and introduces the wide spread biological databanks, their explosion, integration, data warehousing, information retrieval, text mining, text repositories for biological research publications, domain specific search engines, web mining, biological networks and visualization, ontology and systems biology. This chapter also illustrates some technical jargon with picture analogy for a novice learner to understand the concepts clearly.

INTRODUCTION

As stated by John Naisbitt, "We are drowning in information but starved for knowledge." There is a sea change in science due to the tsunami of sequences sweeping over databases. The genome sequencing projects are producing the vast amount of data challenging scientists to potentially reveal the structure and functional relationships of genes and proteins. This phenomenal growth of biological data (biodata) has been witnessed owing to genomic and proteomic technologies such as high-throughput sequencing (HTS) and mass spectrometry, genome-wide two-hybrid

DOI: 10.4018/978-1-5225-1871-6.ch007

screening, DNA microarray, etc. The tremendous amount of sequence information accumulating worldwide have been organized in the form of databanks. However, the data present in the databanks is heterogeneous in nature, such as genomic data, DNA, RNA and protein sequences, protein structures, sequence patterns, sequence annotations, metabolomic data, gene expression data, protein-protein interactions, cross-references, etc. Most of these data are available only on the World Wide Web (Etzioni, 1996). The explosion of these data is impossible to print because it is voluminous and moreover the information in it can only be acquired and assimilated with the aid of data mining tools. With the advent of computers with huge storage capacity and enhanced processing speed, it became easier to use and share information in an electronic format. The digital revolution has made digitized information easy to capture, process, store, distribute and transmit (Fayyad & Uthurusamy, 1996; Inmon, 1996). Besides, the internet allows rapid data transfer and sharing across the globe in an inexpensive manner. The emergence of data mining research such as cluster analysis, outlier analysis, pattern analysis, data visualization and analysis tools contributes to the development of more efficient and scalable methods for knowledge discovery in databanks (Wang et al., 2005). Data mining is a natural evolution of information technology along the path of data collection, database creation, database management, and data analysis and interpretation (Han & Kamber, 2001) in a new field of study known as Bioinformatics. Bioinformatics addresses problems related to the storage, retrieval, and analysis of information about the biological sequence, structure, and function (Altman, 1998).Thc main idea of biodata mining is to discover knowledge out of sequence, structure and functional information from the web.

Since structural biology has an enormous impact on our understanding of biology and medicine, some of the examples used in this chapter are from EMBL-EBI's Macromolecular Structure Database (MSD; www.ebi.ac.uk/msd). Recently, the MSD group has been changed its name to the Protein Databank in Europe (PDBe; http://www.ebi.ac.uk/pdbe/), to reflect its close partnership with the wwPDB project. The services and tools name have been changed, but the EBI maintain all existing URLs for external references to the MSD resource. For the 40^{th}anniversary of PDB (2011), PDBe has turned its attention to a focus on the fundamental problem faced by the structural biology community: "how to make the wealth of structural data available to the larger biomedical community?" (Velankar et al. 2011; Golovin et al. 2004)

BACKGROUND

Information Retrieval

Users approach large information spaces like the Web with different motives, to search for a specific piece of information, or to gain familiarity with some general topic or domain, or to navigate something appealing to them. Usually, the needs and preferences of the users are of varying interest. When they navigate through large web structures, they frequently miss their goal of inquiry. Information retrieval uses the Web (and digital libraries) to access information repositories consisting of mixed media and metadata. Information retrieval based on content implies some amount of summarization or compression (Baeza-Yates & Ribeiro-Neto, 1999). The user can give a query so that the information system retrieve relevant documents related to the given query. Further indexing techniques can be used for effective retrieval. Information retrieval can be made efficient by utilizing data mining tools to infer new biological knowledge from existing data (Figure 1).

There are four main components of information retrieval (Manning, Raghvan & Schutze, 2008). They are as follows:

1. **Indexing:** Generates a representation of the document.
2. **Querying:** User preferences are expressed in natural language with logical operators.
3. **Evaluation:** Match user query and document representation.
4. **User Profile Generation:** Record the user preferences to enhance better user retrieval during future access.

There is a thin line difference between text mining and data mining they usually differ in their inputs (Weiss et al. 2004). Text mining methods usually work on

Figure 1. Process of biological data mining

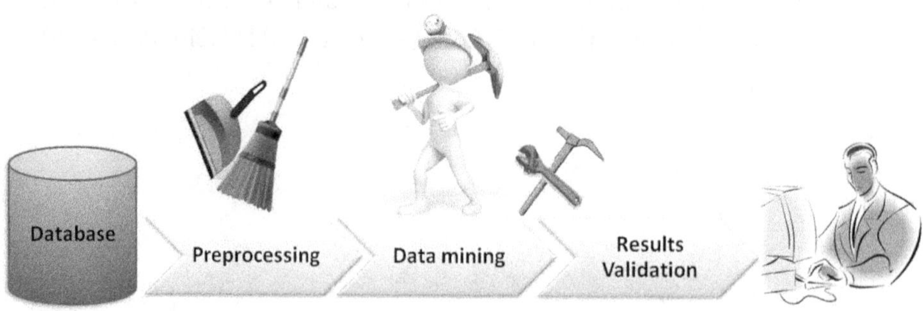

collections of documents which are in human readable format. On the other hand, data mining methods require extensive data preparation for highly structured data format. Despite this difference, the techniques used are common to each other. In general, information extraction methods such as natural language processing or some simple preprocessing are employed by text mining techniques to extract data from texts. Then data mining algorithms are applied to the extracted data (Fiori, 2010).

Knowledge Discovery and Data Mining

The subject of Knowledge Discovery in Databases (KDD) is evolving from the convergence of techniques from multiple disciplines such as databases technology, machine learning, pattern recognition, statistics, information theory, artificial intelligence, data visualization, high-performance computing, etc., (Figure 2). KDD systems incorporate theories, algorithms, and methods from all these fields.

To discover knowledge, the data obtained from different sources "databases" must be organized and archived "data warehousing." Then it should be cleaned and pre-processed, "data cleaning," as it is known that "the garbage in is garbage out," the data quality is important. The core component of KDD is the data mining process that analyses patterns and regularities of the dataset.

Data mining also overlaps with machine learning, statistics, artificial intelligence, databases, and visualization (Figure 2). However, the stress is more on the scalability of the number of features and instances, algorithms and architectures and automation

Figure 2. Knowledge discovery and data mining

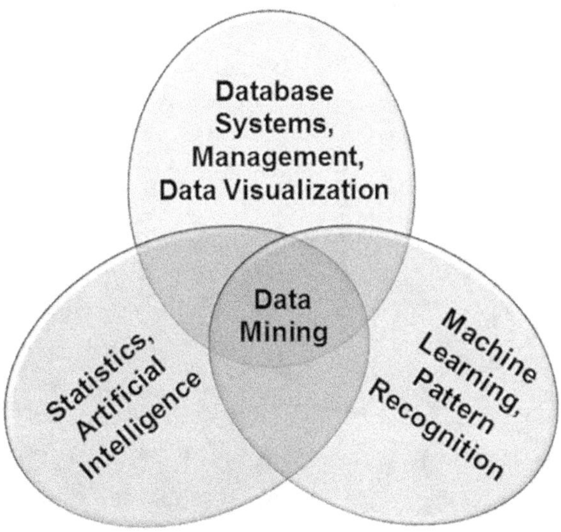

for handling large heterogeneous data. KDD is defined as "the nontrivial process of identifying valid, novel, potentially useful, and ultimately understandable patterns in data." The overall process consists of turning low-level data into high-level knowledge. (Fayyad et al., 1996; Cios, Pedrycz, & Swiniarski, 1998)

Every day humans browse through the web to satisfy their "primitive" need for information, as we humans are perfectly characterized as a species of *Informavores* who "have gained an adaptive advantage because they are hungry for further information about the world they inhabit (and about themselves)" (Dennett, 1991). We humans actively seek, gather, share and consume information to a degree un-approached by other organisms. The evolution of informavores is depicted in Figure 3, to represent various stages of KDD, the Darwin's theory of natural selection may be matched with data selection, transformation of the physical forms can be considered as transformation of data, similarly tool development and transitions from Stone Age to Bronze Age to Iron Age, that are compared analogous to data mining techniques that are used to interpret, validate and analyze data for obtaining knowledge.

The evolution of data mining techniques began when data were first stored on computers. It is a product emerged from research on improvements in data access. This allows *informavores* to navigate through real-time data. Data mining takes this evolutionary process beyond retrospective data access and navigation to prospective and proactive information delivery.

The knowledge discovery process can be divided into six stages (Fayyad, Shapiro, & Smyth, 1996; Adriaans & Zantinge, 1996):

Figure 3. Origin of Informivores

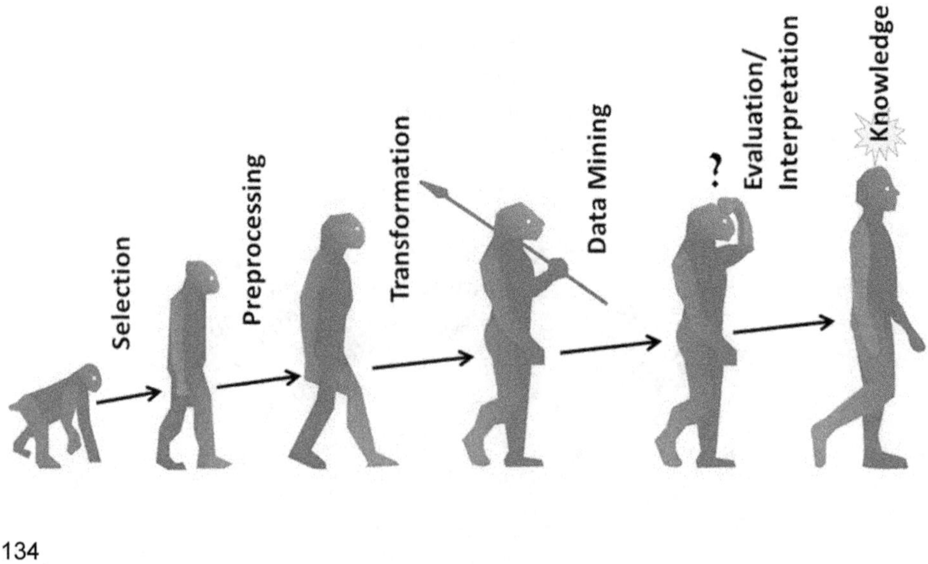

1. **Understanding the Application Domain:** This step involves domain-specific understanding and applications.
2. **Creating the Target Dataset:** This step usually queries the available dataset to create a subset and uses feature ranking and selection techniques to filter a subset of attributes and data points.
3. **Data Preprocessing:** This improves the quality of the raw data for mining. This reduces the time duration and hence increases the mining efficiency. Data preprocessing involves the following
    a. **Data Cleaning:** Data from public repositories such as sequence databanks often contains erroneous and incomplete entries due to sequencing errors. Such low-quality and inconsistent dataset need to be cleaned before data mining. An essential element of this process is to produce a non-redundant dataset. Furthermore, some basic operations are performed such as normalization, noise removal and dealing with missing data, reduction of redundancy, etc.
    b. **Data Integration:** This operation is integral to KDD that integrate information from multiple and heterogeneous datasets. The additional information that is integrated into the available data is referred as data enrichment. For example, in PDBe during "transformation" from the deposition database to the search database, additional derived data are added such as characterization of ligand binding sites, derivation of secondary structure information, mapping data onto other databases such as SwissProt and SCOP, etc.
    c. **Data Dimensionality Reduction and Projection:** This is also termed as 'coding' that finds useful features to represent the data and also reduces the dimensionality of the data. Feature Discretization, and feature extraction techniques can be used to transformed data into appropriate forms for data mining. In the case of Protein structures, this might be the transformation of the protein-binding affinity into groups such as 'non-binder', 'weak binder,' 'moderate binder' and 'strong binder.' In PDBe, PDBeChem can be used to search for ligands, ligand-binding sites, and their geometry.
4. **Data Mining:** Data mining often involves repeated iterative application of particular data mining methods that constitutes one or more of the following functions, such as classification, regression, or clustering.
5. **Interpretation:** This involves interpreting the discovered patterns, as well as extracting low-dimensional information visualization and recognizing patterns in raw data. As humans are good in pattern recognition and understand visual representation, visualization plays an important role in increasing the visual perception of humans. Visualization and knowledge representation techniques

are used to present the mined knowledge to the user. It is also used to evaluate the mined patterns automatically or semi-automatically to identify true or interesting patterns for the user. For example, PDBeallows the user to visualize the structure using AstexViewer.

6. **Use of Discovered Knowledge:** The final step extracts the discovered knowledge and incorporates into the performance system for documenting and reporting. The accuracy of the recorded data must not be overlooked as it depends on domain specific knowledge that assists with the subjective analysis of results. This step may also include taking actions such as checking and resolving potential conflicts with previously believed knowledge. Much attention has been given to the data mining phase. However, pre-processing steps such as data cleaning play a significant role in the validity of the results.

Data Mining

The core component of KDD is data mining, and the data mining technology has revamped for decades, especially in research areas such as bioinformatics, artificial intelligence, and machine learning (Brachmanand Anand, 1996). The maturity of these technologies coupled with high-performance relational database engines brings these technologies for practical applications for data warehousing. Decision support systems (DSS), executive information systems (EIS) and query/report writing tools are used to produce reports about data. Another use of these tools is to detect trends and patterns in the data that will help answer some questions about the problem.

In general, a query is formulated to access the validity of the results. Once the knowledge is discovered, it is verified for the existence of patterns that can answer the queries. This is referred as verification mode. In this, a user of a DSS generates a hypothesis about the data, tests the hypothesis by querying against the data and verifies the results for the validity of the hypothesis. If the response is affirmatory, the process ends. However, if the response is non-affirmatory, a revised query is formulated for an iterative process till the user finds it valid or not (Alexis & Mathews, 1999).

In general, data mining distinguishes patterns from large amounts of unprocessed data that are stored in relational databases or information repositories such as flat text files. Flat files are frequently used for biologists because of its simplicity. Data mining is a multidisciplinary field of interest (Figure 3) that integrates ideas of database management systems, data visualization, machine learning, neural networks, statistics, pattern recognition, signal processing, etc. (Wong, 2002). Data mining tasks can be categorized into two, descriptive and predictive (Han & Kamber, 2000). Descriptive data mining describes the data by discovering interesting patterns or relationships. Whereas predictive data mining classifies the behavior of the model based on the available data. For example, in PDBeMine, a mart-based system that allows you to perform sophisticated searches using any of the 90 entities available in the database, organized into eight different marts. The user can fine tune the settings by adjusting more than 1500 attributes, and use these attributes in combination to make complex searches of more than 500 entity relations. Rather than returning a list of PDB entries, PDBeMine returns only the entity records and attributes that the user is interested, obviating the need to plow through individual records to find the data that you want (Tagari, et al., 2006).

The six main stages of data mining (Han & Kamber, 2001; Mitra, Pal, &Mitra, 2002) are as follows:

1. **Classification:** This model maps or classifies a data item into one of the several predefined classes. The model is derived by analyzing a set of training data that have been explicitly labeled with the classes that they belong to. Then the model is used to predict the test set of class objects whose class label is not known. For example, PDBefold allows you to perform pairwise or multiple comparisons and 3D alignments of structures
2. **Regression:** The purpose of this model function is to map a data item to a real-valued prediction variable.
3. **Clustering:** This function maps a data item into one of the several clusters, in which the classes are determined from the data. This is in contrast to classification, in which the classes are predetermined, cluster analysis is used in situations where the training data do not have any known class labels. The purpose of clustering is to generate class labels for the data. The data objects are typically clustered so that the objects within a cluster are highly similarity to each other and they are dissimilar to objects in other clusters or outliers. For example, in the PDBe database groups represent a family by structure & sequence. Grouping is done to isolate common folds of diverse protein families, in which "Lysozyme" is the most common fold pattern. Unique folds are outliers.
4. **Association Rule Mining:** This generates rules from the data and associate relationships that frequently occur in the dataset, such as co-expression of genes. It is not difficult to develop programs for detecting associations in a large database. However, such a program may return many associations that it is practically hard to differentiate interesting patterns or associations from inappropriate data. For example, in the PDBe database, 30% protein structures in PDBe have associated ligands.
5. **Summarization or Condensation:** This generates a concise description on a subset of data. For example, PDBeView Atlas pages provide an overview of

an individual PDB entry in a user-friendly layout and serve as a starting point for further browsing, searching or analysis (Velankar, et al., 2011a,b).

6. **Deviation Detection:** The purpose of the model is to model the states of data, generating the sequence or to extract and report deviation and trends over time. Sequential pattern discovery has been an active research area in bioinformatics and is applied to gene sequences, time-series analysis, etc. There are many programs that are available for this purpose. The PDBeMotif service provides a pattern (motif) analysis of sequence, structure, and interactions of ligands and their binding properties.

There are two types of learning that arise from the dataset; they are directed or supervised learning, and undirected or unsupervised learning (Kuonen, 2003). In the case of supervised learning, the objective is to build a data model that describes a particular variable of interest and represents the available dataset. Whereas in unsupervised learning, none of the variables is separated out as a target, the purpose is to establish some relationship among all the variables. This modeling finds patterns or similarities among groups of records without the use of predefined classes. For example, association rule mining, clustering, summarization, and visualization are typical to unsupervised learning. In protein modeling, the search for a template to build a homologous structure is analogous to supervised learning. Whereas in fold-based threading, which lacks sequence homology (unsupervised) and hence searches for all possible folds in a database without considering predefined cutoff scores for sequence homology. This is analogous to unsupervised learning.

BIOLOGICAL DATABANKS

A database is defined as a 'systematic collection of information.' It is an organized way of storing enormous information accumulating worldwide especially due to DNA sequencing technologies. Such databases are designed to manage a large amount of homogeneous structured data that is shared among distributed users. Biological databases are publicly accessible to the users and are usually redundant. To address the problem of redundancy in the nucleotide sequences, GenBank, EMBL and DDBJ established the agreement for data redistribution (Brunak, et al., 2002) and formed International Nucleotide Sequence Database Collaboration. The International Nucleotide Sequence Database Collaboration (INSDC) at Heidelberg, Germany, shares sequence information among three major sequence repositories: In North America, the National Center for Biological Information (NCBI), a division of National Library of Medicine (NLM), at the National Institute of Health (NIH), Bethesda, MD has GenBank and GenPept. Also Georgetown University's National

Biomedical Research Foundation (NBRF) Protein Identification Resource (PIR). In Europe, the European Molecular Biology Laboratory (EMBL), European Bioinformatics Institute (EBI) at Heidelberg, Germany maintains nucleotide sequence database. The Swiss Institute of Bioinformatics (SIB)'s Expert Protein Analysis System (ExPASy) maintains Swiss-Prot & TrEMBL protein sequence databases. In Asia, the National Institute of Genetics (NIG) supports the Center for Information Biology's DNA Data Bank of Japan (DDBJ) at Shizuoka prefecture. These three organizations exchange data on a daily basis. Any nucleotide sequence submitted to any of the three major repositories is shared by other major repositories on a daily basis. They largely 'mirror' one another and share accession codes. However, each of these databases has their specific format.

There are approximately 126,551,501,141 bases in 135,440,924 sequence records in the traditional GenBank divisions and 191,401,393,188 bases in 62,715,288 sequence records in the WGS division as of April 2011 (Benson, Karsch-Mizrachi, Lipman, Ostell, & Sayers, 2011). Like an E.coli that doubles every 30 minutes, the size of GenBank is doubling every 18 months. The exponential growth of information can be explained by Moore's law, according to this law" the number of transistors in a single microchip is doubled every 18 months". This can be correlated with data and information explosion, the size and number of databases are growing exponentially. The number of accesses to Swiss-Prot has grown by approximately one million added connections per year(Yi-Ping Phoebe, 2005). The UniProtKB/Swiss-Prot release of Sep 2011 contains 532146 sequence entries, comprising 188719038 amino acids abstracted from 201639 references and of UniProtKB/TrEMBL contains 16886838 sequence entries, comprising 5477504111 amino acids. Discovery of knowledge from this huge wealth is really a challenging task. How to mine information from overwhelming data?, is the question of extreme importance.

All the existing molecular biology databases can be divided into two categories namely, primary and secondary databases. The former is to store sequence data derived directly from the experimental characterization of nucleic acid or proteins. Thus GenBank, EMBL, DDBJ and the Protein Data Bank (PDB) are primary databases. Secondary databases are specialized databases which derive data from the primary databases. There are about 1897 databases developed by various bioinformatics scientists into different category. Based on the functional characterization the existing databases can be classified into the following categories (Table 1).

A catalog of bioinformatics databases is listed in the annual database issue of Nucleic Acids Research (Galperin & Cochrane, 2011). The18th Annual Database Issue (2011) of NAR features descriptions of 96 new online databases covering a variety of molecular biology data and 83 data resources that have previously been published in NAR or other journals. The accompanying NAR online Molecular Biology Database Collection (http://www.oxfordjournals.org/nar/database/a/) now

Table 1. Different categories of databases in bioinformatics

Sr. No.	Category of Databases
1	Nucleotide Sequence Databases
2	RNA sequence databases
3	Protein sequence databases
4	Structure Databases
5	Genomics Databases (non-vertebrate)
6	Metabolic and Signaling Pathways
7	Human and other Vertebrate Genomes
8	Human Genes and Diseases
9	Microarray Data and other Gene Expression Databases
10	Proteomics Resources
11	Other Molecular Biology Databases
12	Organelle databases
13	Plant databases
14	Immunological databases
15	Cell biology

Source: Nucleic Acids Research Journal – Database Issue.

includes 1330 data sources. Further, in 2016, description of 62 more databases has been included which now makes it a total of 1685 databases (Rigden, Fernandez-Suarez & Galperin 2016).

Sequence entries are described in several different formats in GenBank, Swiss-Prot, and EMBL. GenBank developed the ASN.1 (Abstract Syntax Notation One) format while Swiss-Prot has its own format, likely the EMBL. The introduction of XML as the data exchange format has also given rise to several variants of the XML representations of bioinformatics data. Besides ASN.1, GenBank uses GB see XML (weblink ref) format and enable the conversion of ASN.1 data to GB see XML. Swiss-Prot's XML format is termed as SPTr-XML (www.ebi.ac.uk/swissprot/sp-ml/mapping-guide.html) and the EMBL has developed the XEMBL (www.ebi.ac.uk/emblem). In spite of various data formats, a unified protocol for data exchange has not been established and hence leads to complications in managing bioinformatics data (Brunak, et al., 2002). Each of these databases provides a flat file of data. However, researchers have diversified interest retrieve information from various databases to support their experiments, hypothesis, and interpretation. Hence, the concept of data warehousing is applied in bioinformatics databases. This provides a better solution to manage different views of data and ensures data interoperability. An integrated information-retrieval system in NCBI is illustrated in Figure 4, a search interface on the top layer (i.e. Entrez) that connects various data sources is not shown. This provides a network of pre-computed links between the key online molecular biology databases. The popular integrated systems in NCBI facilitate access to more than twenty interlinked databases.

The establishment of the data warehouse in bioinformatics greatly facilitates biological research for high-level data analysis, abstraction, and extraction of new knowledge. "A biological data warehouse is a subject-oriented, integrated, non-volatile, expert interpreted collection of data in support of biological data analyses and knowledge discovery" (Schönbach, Kowalski-Saunders, & Brusic, 2000).

DATA WAREHOUSING

In the data-driven field of bioinformatics, data warehouses have emerged to facilitate biological data analysis. The processing steps such as data cleaning, data integration, and analysis are commonly used in data warehousing. In bioinformatics, data warehousing is an effort, which help biologists to plan the design of critical experiments for their research. This includes data collection from various databases, resolving

Figure 4. Integrated information-retrieval system in NCBI

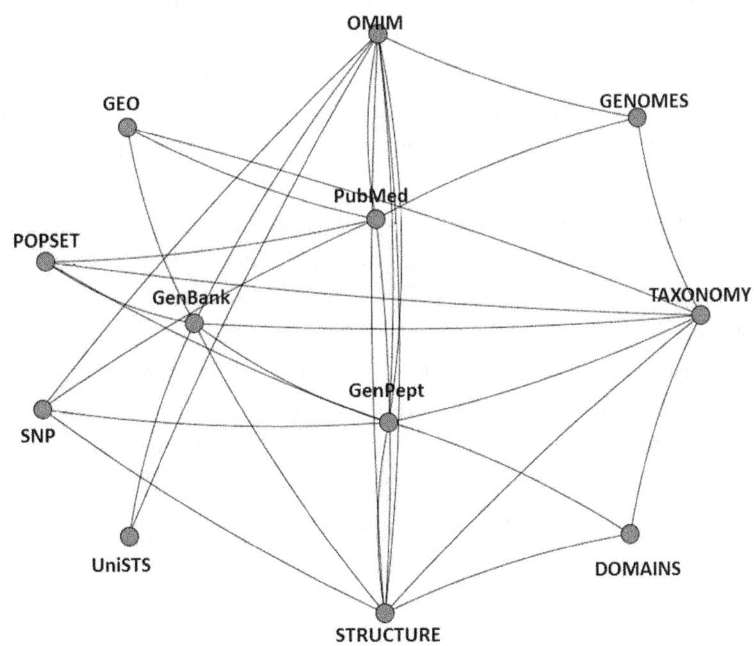

conflicts, recognize patterns and transform data into a form usable for knowledge discovery. This requires effective storage, integration, and organization of a large volume of data into a single, well-structured repository suitable for analytical processing (Reddy et al., 2010). Biological data warehousing has specific requirements depending on the nature of biological data and its applications. By using the general principles of data warehousing, we can formulate domain-specific requirements, so as to make an efficient biological data analysis platform using biological warehouse. This will facilitate data analysis using data mining techniques and contribute to the discovery of new knowledge for biological research. The application of data warehousing principles to bioinformatics data is incredible. There are few examples in biological data, such as gene expression data warehouse (Yi-Ping Phoebe, 2005) and genomic data warehouse (Cornell, et al., 2003).

There are two data integration approaches, virtual or materialized integration (Durand, et al., 2003). Bioinformatics data adopt either of this approach for federated databases that provides a software interface to multiple data sources that are maintained independently. Such acts between the virtual warehouse and the data sources that contain physical data. The examples include DiscoveryLink (Haas, 2001), Kleisli (Chung & Wong, 1999), SRS (Zdobnov, Lopez, Apweiler, & Etzold, 2002), Entrez (Wheeler, et al., 2003) and TAMBIS (Stevens, et al., 2000). There are annotation tools that are essential for data enrichment that aids structural and functional annotation. These tools facilitate data preprocessing either automatically or with human intervention. When bioinformatics data adopt materialized data integration, it should consider the amount of data storage space required. The technologies used for integrating biological data are summarized by Wong (2002). An open source toolkit, BioWarehouse was introduced by Lee et al. (2006)for constructing bioinformatics database warehouses using MySQL and Oracle relational database managers. BioWarehouse connects all database components into a common husband maintains a single database management system. It enables multi-database queries using Structured Query Language (SQL) and also facilitates comparative analysis and data mining. BioWarehouse supports the integration of pathway-centric databases including ENZYME, KEGG, and BioCyc in addition toUniProt, GenBank, NCBI Taxonomy, CMR databases, and Gene Ontology. The loader tools in BioWarehouse were written in the C and JAVA languages; they apply a degree of semantic normalization to their respective source data to decrease semantic heterogeneity. It supports a diverse data types for chemical compounds, biochemical reactions, metabolic pathways, proteins, genes, nucleic acid sequences, organisms, taxonomies, and controlled vocabularies

TEXT MINING

The text is a frequently used among the other multimedia data types. It is a preferred choice for information exchange among common public throughout the world. Biological databases do prefer plain text (flat files) for sharing information resource across the world. The exponential growth of sequences and easy access to sequence records in biological databases is due to a revolution in sequencing technology as well as information technology. Many data mining algorithms are in development for easy extraction of desired information from well-structured relational databases and processed data warehouses. For text mining, retrieval techniques have been developed for indexing and searching unstructured or semi-structured text documents from databases. However, these traditional techniques are not sufficient for knowledge discovery and mining from the overwhelming databases. Text mining is an emerging discipline with novel search engines for retrieval of text-based information. Besides traditional data mining principles, modified string matching techniques can be used for pattern search in text or string of characters. String matching is an important aspect of data mining with potential applications in Bioinformatics.

Bioinformatics and big data analysis involve worldwide researchers from many different disciplines, for instance, genetics, molecular biology, biochemistry, medical science, statistics, computer science, etc. It is very important to search the document collections of published papers retrieve background knowledge on specific topics. At the same time, it is cumbersome to do a literature search to get all relevant parts of research papers to understand the relationship of genes and proteins concerning structure, function, clinical disorders, and so on. It is rightly mentioned "to extract and analyze the data perhaps poses a much bigger challenge for researchers than to generate the data" (Holloway, van Laar, Tothill, & Bowtell, 2002). Data integration for inferring new knowledge is essential to many biological studies. It is very difficult to manage the overwhelming medical and biological research publications in the text repositories eg., PubMed Central (http://www.pubmedcentral.nih.gov/).

PUBMED CENTRAL

NCBI supports a free digital repository for unrestricted access to electronic biomedical and life sciences journal collection. It is to be noted that the PubMed Central (PMC) is a digital library and not a journal publisher, which offers access to many journals. PMC maintains a single repository and stores data from diverse sources. This also simplifies the searching task and integrates literature with other resources Figure 4. *Entrez* information retrieval system makes easy access to PMC for citations and abstracts of biomedical text.

An FTP service is available at PMC to download the source files for any article in the PMC Open Access Subset. The source files can be any of the following, '.nxml file', i.e., an XML version of a full text article, a PDF file of the article, images used in the article, graphics for mathematical equations or chemical reactions and supplementary research data or videos.

DOMAIN SPECIFIC SEARCH ENGINES

Domain-specific search engines perform searches in specialized repositories using specific keywords usually exploits techniques used for general purpose search engines and apply it for a narrower domain. One such specific project for biomedical articles is Information Hyperlinked over Proteins, iHop (Hoffmann & Valencia, 2004). This project categorizes biomedical text for genes and proteins and provides instant access to the documents for certain gene or group of genes. The iHop project uses MeSH Headings, MeSH stands for Medical Subject Headings (http://www.nlm.nih.gov/mesh/) for effective web mining. NLM's MeSH celebrated its golden jubilee on 2010.

MesH terms are assigned to topics (indexing) of every publication in Medline. This permits searches at varying levels of specificity, improves appropriate results and restrict irrelevant citations (Fiori, 2010). Another interesting feature iHop applies a combination of natural language parsing and gene identifiers to extract information from PubMed abstracts and MeSH headings to get an overview of gene functionalities. However, there is a limitation; gene functionalities may not be derived based on abstracts and headings alone. It has been reported in research publications that iHop missed statements that deals with protein-protein interactions and extracts statements related to the gene or protein. Moreover, a physician's evaluation of randomly-selected papers (Reeve, Han, & Brooks, 2007) shows that the author's abstract does not always reflect the entire contents of the full text and the crucial biological information. For these reasons working only on the abstracts is a critical limitation. Considering its strengths and this limitation iHop is a very powerful domain specific search engine, but it still requires a lot of human intervention to fulfill the needs of inferring knowledge and providing biological validation. Considering the Web as a huge repository of distributed hypertext, the results from text mining have great influence in Web mining and information retrieval (Mitra & Acharya, 2003).

WEB MINING

World Wide Web (WWW) can be considered as the largest distributed database that ever existed. However, at the same time, it is also considered to be the most disorganized database as well. Therefore, mining of the The Web is challenging, as it comprises of semi-structured (HTML, XML), hyperlink and dynamic information that are constantly generated. The Web provides wide accessibility to the users for heterogeneous information, typically unlabeled, distributed, semi-structured, time-varying, and high-dimensional information. It provides greater access to networks and free access to a large publishing medium. These features allow people to access the Web continuously for knowledge discovery.

Mining relevant bioinformation from the web is really a challenging task to discover knowledge. However, some human interference is also needed to fine tune the information. Web mining refers to the use of data mining techniques to automatically retrieve, extract, and evaluate information for knowledge discovery. The major components of Web mining (Kohavi, Masand, Spilipoulou, &Srivastava, 2002) include the following:

1. **Information Retrieval:** Refers to the automatic retrieval of relevant documents, using document indexing and search engines.
2. **Information Extraction:** Helps to identify the pieces of information that constitute the semantic core of the Web.
3. **Generalization:** Relates to aspects from pattern recognition or machine learning, and it utilizes clustering and association rule mining.
4. **Analysis:** Corresponds to the extraction, interpretation, validation, and visualization of the knowledge obtained from the Web.

In today's data processing environment, most of the text data is stored in a compressed form similar to a simple word compression/ abbreviation (eg., Deoxyribonucleic Acid as 'DNA'). Usually, classical text compression algorithms, such as the Lempel-Ziv family of algorithms, are used to compress text databases. Hence access to text information in the compressed domain will become a challenge shortly. Other established mathematical principles for data reduction have also been applied in text mining to improve the efficiency of these systems. One such technique is the application of principal component analysis (Mitra & Acharya, 2003).

BIOLOGICAL NETWORKS AND VISUALIZATION

The biomolecular interactions/ network of a cell can be represented using graphs. A set of connected molecular interactions can be considered as a pathway. The cellular system involves complex interactions between proteins, DNA, RNA, and smaller molecules and can be categorized in three broad subsystems namely, metabolic network or pathway, protein network, and gene regulatory network. For well-studied organisms, especially microbes such as E. coli, a composite collection of metabolic reactions are organized into large online databases, such as EcoCyc (Karp et al., 2002). The analysis of large-scale gene expression can be conceptualized as a genetic feedback network. The ultimate goal of microarray analysis is the complete reverse engineering of the genetic network (Wang, Zaki, Toivonen, &Shasha, 2005).

Visualization

There are a number of generic visualization software products such as AVS, IBM Visualization Data Explorer, SGI Explorer, Visage, Khoros, SAGE, SDM, S-Plus, SPSS, SciAn, MatLab, Mathematica, MAPLE, NetMap, etc. Visualization tools are composed of the following:

- Visualization techniques classified based on tasks, data structure, or display dimensions.
- Visual perception type, e.g., selection of graphical primitives, attributes, attribute resolution, the use of color in fusing primitives.
- Display techniques, e.g., static or dynamic interactions; representing data as line, surface or volume geometries; showing symbolic data as pixels, icons, arrays or graphs (Fayyad, Grinstein, & Wierse, 2001).

The protein network shown in Figure 5 is generated and visualized using STRING (Jensen et al., 2008). STRING is a database of known and predicted protein interactions. The interactions include direct (physical) and indirect (functional) associations; they are derived from four sources: genomic context, high-throughput experiments, co-expression and previous knowledge (text mining from literature).

Figure 5. Protein networks

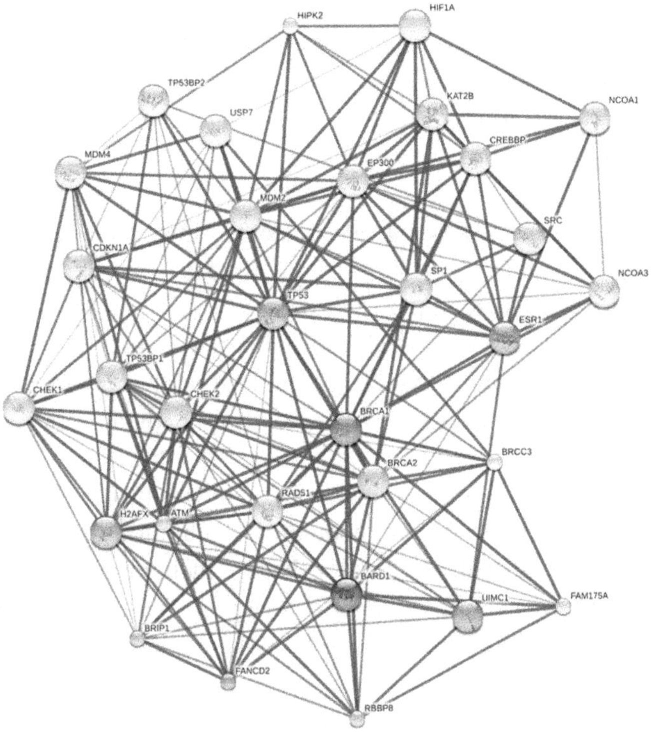

FUTURE RESEARCH DIRECTIONS

Ontology

The upcoming generation of the World Wide Web, the so-called semantic Web, aims at improving the "semantic awareness" of computers connected via the Internet. The Semantic Web requires that information is given a well-defined meaning through a machine-processable representation of the world, often referred to as an ontology. Ontologies are the structural frameworks for sharing domain specific interest. Such unifying conceptual framework promotes communications among people, interoperability among systems, reliability, reusability and specification and so on. Ontologies are widely used in artificial intelligence, the semantic web, software engineering, systems engineering, biomedical informatics, etc. Bio ontologies are in development.

Systems Biology

Systems Biology deals with a system-level understanding of structure and dynamics of genes and proteins (Ideker, Galitski, & Hood, 2001). According to the prominent systems biologist, Kitano (2002a), a biological system can be understood at systems level with an insight into four key properties:

1. **Systems Structures:** These include molecular networks, biochemical pathways as well as and interactions of genes that modulate the physical properties of intracellular and multicellular structures.
2. **Systems Dynamics:** This deal with the behavior of the system over time under various conditions that can be understood through metabolic analysis, sensitivity analysis and other dynamic analysis methods such as phase portrait and bifurcation analysis.
3. **Systems Control:** This deal with the mechanisms that systematically control the state of the cell that can be modulated to minimize malfunctions and provide potential therapeutic targets for the treatment of disease.
4. **Systems Design:** This deal with the design principle and simulation strategies to design and modify biological systems that are having desired properties.

There are two distinct branches of computational biology:

1. Knowledge discovery, or data mining that extracts patterns from volumes of experimental data and frame hypotheses as a result
2. Simulation-based analysis that tests hypotheses with in-silico experiments, providing predictions to be tested by in vitro and in vivo studies (Kitano, 2002b).

To overcome different data formats used by research groups for data exchange, some special markup languages such as Systems Biology Markup Language (SBML) (Hucka et al., 2003), CellML (http://www.cellml.org/), and the Systems Biology Workbench can be used as a de facto standard for the research groups to exchange their models or create commonly accepted repositories. This provides an open software platform for modeling and analysis of next-generation databases that are concerned with biological pathways such as Kyoto Encyclopedia of Genes and Genomes (KEGG), Alliance for Cellular Signaling (AfCS), and Signal Transduction Knowledge Environment (STKE), by enabling them to develop machine executable models rather than merely human-readable forms (Kitano, 2002b).

Building a full-scale organism model or even a whole-cell or organ model is a challenging task and require further research. Many research groups such as Virtual Cell (Schaff, Slepchenko, & Loew, 2000) and E-Cell (Tomita et al., 1999) have started working on this direction. Integrations of biological networks and processes predict models that can be corroborated with clinical data. Integrating heterogeneous simulation models require integration of data of multiple scales, resolutions, and modalities (Wang, et al., 2005).

CONCLUSION

Biological data mining is an important area for fishing knowledge from the ocean of information. In the recent past, a lot of implementations have happened on large databases involving association rules, classification, clustering, text document retrieval, outlier analysis, etc. However, the field is open for research, especially on biological databases. The research communities are also facing challenging issues in this field. This chapter provides a brief overview of biological data mining and an introduction to knowledge discovery from databases. The data mining strategies have been described for novice biologists entered into the field of bioinformatics. Currently, the bioinformatics databases have undergone a drastic change in the search strategies by the implementation of data mining concepts as explored from PDBe. The concepts of data mining and bioinformatics have been co-evolved and presently both the field interacts and collaborates actively with each other. The collaboration between these two fields has just started, and many more fruitful results will appear shortly.

REFERENCES

Adriaans, P., & Zantinge, D. (1996). *Data Mining*. Harlow, UK: Addison-Wesley Longman.

Alexis, L., & Mathews, L. (1999). *Fundamentals of Information Technology*. Leon Press.

Altman, R. A. (1998). Curriculum in Bioinformatics: The time is ripe. *Bioinformatics (Oxford, England)*, *14*(7), 549–550. doi:10.1093/bioinformatics/14.7.549 PMID:9841111

Baeza-Yates, R., & Ribeiro-Neto, B. (1999). *Modern Information Retrieval*. Reading, MA: Addison-Wesley.

Benson, D. A., Karsch-Mizrachi, I., Lipman, D. J., Ostell, J., & Sayers, E. W. (2011). GenBank. *Nucleic Acids Research, 39*(Database issue), D32–D37. doi:10.1093/nar/gkq1079 PMID:21071399

Brachman, R., & Anand, T. 1996. The Process ofKnowledge Discovery in Databases: A Human-CenteredApproach. In Advances in Knowledge Discovery and Data Mining, (pp. 37–58). Menlo Park, CA: AAAI Press.

Brunak, S., Danchin, A., Hattori, M., Nakamura, H., Shinozaki, K., Matise, T., & Preus, D. (2002). Nucleotide Sequence Database Policies. *Science, 298*(5597), 1333. doi:10.1126/science.298.5597.1333b PMID:12436968

Chung, S. Y., & Wong, L. (1999). Kleisli: A new tool for data integration in biology. *Trends in Biotechnology, 17*(9), 351–355. doi:10.1016/S0167-7799(99)01342-6 PMID:10461180

Cios, K. J., Pedrycz, W., & Swiniarski, R. (1998). *Data Mining Methods for Knowledge Discovery*. Dordrecht: Kluwer. doi:10.1007/978-1-4615-5589-6

Cornell, M., Paton, N. W., Wu, S., Goble, C. A., Miller, C. J., Kirby, P., & Oliver, S. G. et al. (2003). GIMS – an integrated data storage and analysis environment for genomic and functional data. *Yeast (Chichester, England), 15*(15), 1291–1306. doi:10.1002/yea.1047 PMID:14618567

Dennett, D. C. (1991). *Consciousness explained.* Boston, MA: Little, Brown, and Co.

Durand, P., Medigue, C., Morgat, A., Vandenbrouck, Y., Viari, A., & Rechenmann, F. (2003). Integration of data and methods for genome analysis. *Current Opinion in Drug Discovery & Development, 6*, 346–352. PMID:12833667

Etzioni, O. (1996). The World-Wide Web: Quagmire or goldmine? *Communications of the ACM, 39*(11), 65–68. doi:10.1145/240455.240473

Fayyad, U., Grinstein, G., & Wierse, A. (2001). *Information Visualization in Data Mining and Knowledge Discovery*. San Francisco: Morgan Kaufmann Publishers.

Fayyad, U., Shapiro, G. P., & Smyth, P. (1996). The KDD process for extracting useful knowledge from volumes of data. *Communications of the ACM, 39*(11), 27–34. doi:10.1145/240455.240464

Fayyad, U., & Uthurusamy, R. (1996). Data mining and knowledge discovery in databases. *Communications of the ACM, 39*(11), 24–27. doi:10.1145/240455.240463

Fayyad, U. M., Piatetsky-Shapiro, G., Smyth, P., & Uthurusamy, R. (Eds.). (1996). *Advances in Knowledge Discovery and Data Mining*. Menlo Park, CA: AAAI/MIT Press.

Fiori, A. (2010). *Extraction of biological knowledge using data mining techniques* (Doctoral dissertation). Available from database and data mining group, Politecnico Di Torino, XXII cycle, 2010 Ph.D. Thesis. Retrieved from http://dbdmg.polito.it/wordpress/wp-content/uploads/2010/12/Thesis_Fiori_Alessandro.pdf

Galperin, M. Y., & Cochrane, G. R. (2011). The 2011 Nucleic Acids Research Database Issue and the online Molecular Biology Database Collection. *Nucleic Acids Research*, *39*(Database issue), D1–D6. doi:10.1093/nar/gkq1243 PMID:21177655

Golovin, A., Oldfield, T. J., Tate, J. G., Velankar, S., Barton, G. J., Boutselakis, H., & Dimitropoulos, D. (2004). E-MSD: An integrated data resource for bioinformatics. *Nucleic Acids Research*, *32*(Database issue), D211–D216. doi:10.1093/nar/gkh078 PMID:14681397

Haas, L. M., Schwartz, P. M., Kodali, P., Kotlar, E., Rice, J. E., & Swope, W. C. (2001). DiscoveryLink: A system for integrated access to life sciences data sources. *IBM Systems Journal*, *40*(2), 489–511. doi:10.1147/sj.402.0489

Han, J., & Kamber, M. (2001). *Data Mining: Concepts and Techniques*. San Diego, CA: Academic Press.

Han, J., & Kamber, M. (2001). *Data Mining: Concepts and Techniques*. San Francisco, CA: Morgan Kaufmann Publishers.

Hoffmann, R., & Valencia, A. (2004). A gene network for navigating the literature. *Nature Genetics*, *36*(7), 664. http://www.ihop-net.org/ doi:10.1038/ng0704-664 PMID:15226743

Holloway, A., van Laar, R. K., Tothill, R. W., & Bowtell, D. (2002). Options available—from start to finish—for obtaining data from DNA microarrays II. *Nature Genetics*, *32*(Supplement), 481–489. doi:10.1038/ng1030 PMID:12454642

Hucka, M., Finney, A., Sauro, H. M., Bolouri, H., Doyle, J. C., Kitano, H., & Wang, J. (2003). The system biology markup language (SBML): A medium for representation and exchange of biochemical network models. *Bioinformatics (Oxford, England)*, *19*(4), 524–531. doi:10.1093/bioinformatics/btg015 PMID:12611808

Ideker, T., Galitski, T., & Hood, L. (2001). A new approach to decoding life: Systems biology. *Annual Review of Genomics and Human Genetics*, *2*(1), 343–372. doi:10.1146/annurev.genom.2.1.343 PMID:11701654

Inmon, W. H. (1996). The data warehouse and data mining. *Communications of the ACM*, *39*(11), 49–50. doi:10.1145/240455.240470

Jensen, L. J., Kuhn, M., Stark, M., Chaffron, S., Creevey, C., Muller, J., & von Mering, C. et al. (2009). STRING 8- a global view on proteins and their functional interactions in 630 organisms. *Nucleic Acids Research*, *37*(Database issue).

Karp, P., Riley, M., Saier, M., Paulsen, I., Collado-Vides, J., Paley, S., … Gama-Castro, S. (n.d.). The EcoCyc database. Nucleic Acids Research, 30, 56–58.

Kitano, H. (2002a). Systems biology: A brief overview. *Science*, *295*(5560), 1662–1664. doi:10.1126/science.1069492 PMID:11872829

Kitano, H. (2002b). Computational systems biology. *Nature*, *420*(6912), 206–210. doi:10.1038/nature01254 PMID:12432404

Kohavi, R., Masand, B., Spilipoulou, M., & Srivastava, J. (2002). Web mining. *Data Mining and Knowledge Discovery*, *6*(1), 5–8. doi:10.1023/A:1013266218887

Kuonen, D. (2003). Challenges in Bioinformatics for Statistical Data Miners. *Bulletin of the Swiss Statistical Society*, *46*, 10–17.

Lee, T. J., Pouliot, Y., Wagner, V., Gupta, P., Stringer-Calvert, D. W. J., Tenenbaum, J. D., & Karp, P. D. (2006). BioWarehouse: A bioinformatics database warehouse toolkit. *BMC Bioinformatics*, *7*(1), 170. doi:10.1186/1471-2105-7-170 PMID:16556315

Manning, C., Raghavan, P., & Schutze, H. (2008). *Introduction to Information Retrieval*. Cambridge University Press. doi:10.1017/CBO9780511809071

Mitra, S., & Acharya, T. (2003). *Data Mining, Multimedia, soft computing and Bioinformatics*. Hoboken, NJ: John Wiley & Sons, Inc.

Mitra, S., Pal, S. K., & Mitra, P. (2002). Data mining in soft computing framework: A survey. *IEEE Transactions on Neural Networks*, *13*(1), 3–14. doi:10.1109/72.977258 PMID:18244404

Nucleic Acids Research – Database Issue. (n.d.). Available online at https://www.oxfordjournals.org/our_journals/nar/database/cap/

Reddy, G. S., Srinivasu, R., Rao, M. P. C., & Rikkula, S. R. (2010). Data Warehousing, Data Mining, OLAP and OLTP Technologies are Essential Elements to Support Decision-Making Process inIndustries. *International Journal on Computer Science and Engineering*, *2*(9), 2865–2873.

Reeve, L. H., Han, H., & Brooks, A. D. (2007). The use of domain-specific concepts in biomedical text summarization. *Information Processing and Management: An International Journal, 43*(6).

Rigden, D. J., Fernandez, M., & Galperin, Y. (2016). The 2016 database issue of Nucleic Acids Research and an updated molecular biology database collection. *Nucleic Acids Research, 44*(D1), D1–D5. doi:10.1093/nar/gkv1356 PMID:26740669

Schaff, J., Slepchenko, B., & Loew, L. (2000). Physiological modeling with virtual cell framework. *Methods in Enzymology, 321*, 1–23. doi:10.1016/S0076-6879(00)21184-1 PMID:10909048

Schönbach, C., Kowalski-Saunders, P., & Brusic, V. (2000). Data warehousing in molecular biology. *Briefings in Bioinformatics, 1*(2), 190–198. doi:10.1093/bib/1.2.190 PMID:11465030

Stevens, R., Baker, P., Bechhofer, S., Ng, G., Jacoby, A., Paton, N. W., & Brass, A. et al. (2000). TAMBIS: Transparent access to multiple bioinformatics information sources. *Bioinformatics (Oxford, England), 16*(2), 184–185. doi:10.1093/bioinformatics/16.2.184 PMID:10842744

Tagari, M., Tate, J., Swaminathan, G. J., Newman, R., Naim, A., Vranken, W., & Velankar, S. (2006). E-MSD: Improving data deposition and structure quality. *Nucleic Acids Research, 34*(90001), D287–D290. doi:10.1093/nar/gkj163 PMID:16381867

Tomita, M., Hashimoto, K., Takahashi, K., Shimizu, T. S., Matsuzaki, Y., Miyoshi, F., & Hutchison, C. A. et al. (1999). E-cell: A software environment for while-call simulation. *Bioinformatics (Oxford, England), 15*(1), 72–84. doi:10.1093/bioinformatics/15.1.72 PMID:10068694

Velankar, S., Alhroub, Y., Alili, A., Best, C., Boutselakis, H. C., Caboche, S., & Kleywegt, G. J. et al. (2011b). PDBe: Protein Data Bank in Europe. *Nucleic Acids Research, 39*(Database issue), D402–D410. doi:10.1093/nar/gkq985 PMID:21045060

Velankar, S., Best, C., Beuth, B., Boutselakis, C. H., Cobley, N., Sousa Da Silva, A. W., & Kleywegt, G. J. et al. (2011a). PDBe: Protein Data Bank in Europe. *Nucleic Acids Research, 38*(Database issue), D308–D317. PMID:19858099

Wang, J. T. L., Zaki, M. J., Toivonen, H. T. T., & Shasha, D. E. (2005). *Data Mining in Bioinformatics*. London, UK: Springer-Verlag.

Weiss, S., Indurkhya, N., Zhang, T., & Damerau, F. (2004). *Text Mining: Predictive Methods for Analyzing Unstructured Information*. New York: Springer-Verlag.

Wheeler, D. L., Church, D. M., Federhen, S., Lash, A. E., Madden, T. L., Pontius, J. U., & Wagner, L. et al. (2003). Database resources of the National Center for Biotechnology. *Nucleic Acids Research*, *31*(1), 28–33. doi:10.1093/nar/gkg033 PMID:12519941

Wong, L. (2002). Datamining: Discovering Information from Bio-Data. In T. Jiang, Y. Xu, & M.Q. Zhang (Eds.), Current topics in computational molecular biology, (pp. 317-342). Cambridge, MA: MIT Press.

Wong, L. (2002). Technologies for Integrating Biological Data. *Briefings in Bioinformatics*, *3*(4), 389–404. doi:10.1093/bib/3.4.389 PMID:12511067

Yi-Ping Phoebe, C. (Ed.). (2005). *Bioinformatics Technologies*. Berlin: Springer-Verlag.

Zdobnov, E. M., Lopez, R., Apweiler, R., & Etzold, T. (2002). The EBI SRS server-new features. *Bioinformatics (Oxford, England)*, *18*(8), 1149–1150. doi:10.1093/bioinformatics/18.8.1149 PMID:12176845

KEY TERMS AND DEFINITIONS

Biological Databases: Collection of data arranged in a meaningful manner.

Biological Networks: The biomolecular interactions/ network of a cell can be represented using graphs.

Data Mining: Data mining often involves repeated iterative application of particular data mining methods that constitutes one or more of the functions to retrieve meaningful information.

Chapter 8
Information Services to Biomedical Science through Mobile Technology Applications

John Paul Anbu
UNISWA, Swaziland

ABSTRACT

Biomedical science is one filed where huge amount of information is generated, distributed over the internet and a number of software tools are also developed to generate information. The quantum of biomedical data along with the proliferation of new data integration technologies have made it important to adopt smart and fast network tools to access information in bioinformatics. It is important to make researchers in biomedical science aware of systematic approaches to access these information. One avenue to implement this approach is to make the biomedical information available through mobile technology which is still missing. It is heartening to see that there are some mobile initiatives taking place in biomedical sciences which provide handy tools for bioinformatics information seekers to access information. This paper is a review of such tools which will aid the library and information professionals to create information literacy in this field in future.

DOI: 10.4018/978-1-5225-1871-6.ch008

INTRODUCTION

The ICT revolution, especially the advancements in the internet and communication technologies, more specifically in mobile technologies and its applications throughout the world, has brought enormous change amongst the providers of information. The world of information science has left with no option but to join to suit with the mobile users who are "connected to educational opportunities from virtually anywhere, making almost every situation a potential learning environment" (McQuiggan et. al. 2015, p.50). Mobile technology has an upper hand when it comes to accessing information as it provides a dynamic access solution because of its mobility. The rapid growth of mobile technology has greatly supported the users to enhance their research activities since it has the element of being current to the content the users get. Traditional information dissemination avenues like libraries, museums, and datacentres have already started shifting from their traditional Online Public Access Catalogues (OPACs) to mobile specific access applications (Zhou &Ramona, 2011; Barile, 2011; Kroski, n.d.). In recent years, many academic libraries have adopted "Mobile Technologies" to meet the users' expectations (Ram, Anbu & Kataria 2011).

Gartner predicted by 2015 the smartphone and tablet usage will increase to 90% (Gartner 2012). Figure 1 illustrates the growth of mobile phones starting from 2005 onwards in millions. A closer look at the graph shows the steady growth in mobile ownership which has reached the 100% mark and has gone above the mark

Figure 1. Growth of mobile ownership in the past 10 years

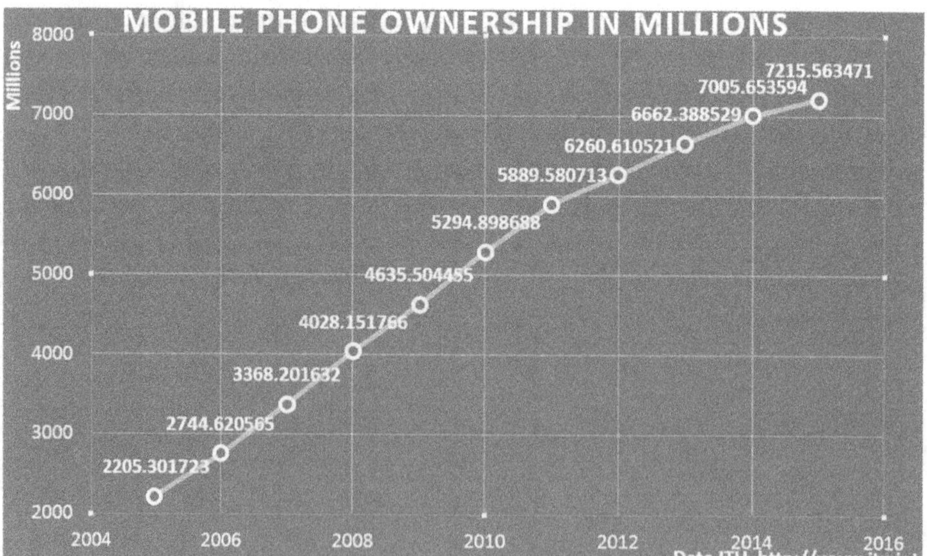

which suggest that multiple mobile gadgets per single user will be very common in the coming years. When it comes to providing alerts and notices, mobile based applications have augmented the traditional support system for better results (Jetty and Anbu, 2013). Mobile technology has been adopted by institutions of higher learning as a media for creating information literacy (Ram, Anbu & Kataria 2014). Wu (2004) believed the future of libraries would be mobile. Mao, Wu, and Huang (2008) introduced a concept and function of the mobile library service.

MOBILE TECHNOLOGY

Mobile technologies include laptops, netbooks, e-readers, tablets, mobile phones, smartphone MP3/ MP4 players and the internet capable handheld devices. In addition, wireless networks, such as 3G, are providing the network infrastructure for users. The proliferation of mobile technology specially the 3G hype, heading towards 4G, and W-Fi innovations have triggered an unprecedented change in the tool for access to the internet. With the 4G and Wi-Fi hype the mobile technology has been used not only for communicating but for transferring enormous amount of data through the network. Specific applications and mobile specific websites have sprung to harness the power of mobile computing. The mobile application platforms started appearing as early as the 1987 by Apple's Newton Helel et. al. (2015). Currently the entry of smartphones into the mobile industry and thc availability of internet dependent applications into the mobile technology has given birth to a number of platforms. As a consequence, major players in the mobile platform industry, Apple's iOS, Google's Android and Microsoft's Windows phone platforms emerged as the forerunners in the mobile platform industry.

Mobile Web and Development Tools

Since the beginning of 1998 when Open Mobile Alliance (OMA) created the first Wireless Application Protocol (WAP) there has been a number of developmental tools for mobile platforms. When the WAP was first derived it was derived to view the content. Over the years, the evolution of further protocols and the need to have applications itself being viewed a drastic change started appearing in the development of mobile web. Basically both the mobile web as well as the desktop web make use of the same technologies such as HTML (Hyper Text Markup Language), CSS (Cascading Style Sheet) and JavaScript. Mobile web applications are nothing different from the desktop applications except that it takes care of the size of the mobile device, its limitations, its mobile specific capabilities and its network capabilities. Device detection, handling of CSS Media Queries, Touch and gesture Events, Data optimised computing and providing location based services are the hallmarks of mobile computing.

There are three kinds of mobile applications environments which are practices in the mobile app world:

1. **Web Applications:** Applications which are built to serve multi platforms irrespective of the operating platforms. Using the browser support of the smart mobile phones mobile web applications are developed using standard HTML coding. Though it cannot be 100% portable to all the different screens and operating environments, it is more reliable and easy to deploy. The need to download them from app stores are minimized and only a link is need to deploy and activate it on the mobiles. JQuery Mobile and Sencha Touch are two most popular mobile web development frameworks which are commonly used in mobile web application development.
2. **Native Applications:** Makes use of the inbuilt mobile specific features and the specific resolutions and screen sizes of which are native to those mobile gadgets. Since it is native to the operating environment as well as to the device, more specific APIs are being used from the native environment to accelerate the application process. Normally a native application can be downloaded only from the specific application store which gives more control over the in-house monetary processes and also the proper execution of services like push notifications and other mobile specific activities.
3. **Hybrid Applications:** Developed using Native application method and within which the web application links are imbedded so that a parallel cross platform actions are controlled. Hybrid applications are built with a combination of web technologies like HTML, JavaScript, CSS and they are hosted inside the native applications so that it can use the mobile browsing environment. Adobe PhoneGap is an open source hybrid application development environment which is used for creating hybrid applications.

According to Gartner and Statistica the global mobile application download has been tremendous and by the end of 2016, it will reach 268.7b from 64.0b in 2012 (Figure 2). The progressing growth trend illustrates the use of apps through mobile devices.

SOFTWARE DEVELOPMENT IN BIOINFORMATICS

Bioinformatics is about using computational approaches to analyze large scale biological data that would be impossible or nonviable to approach with other meth-

Figure 2. Downloads of mobile apps

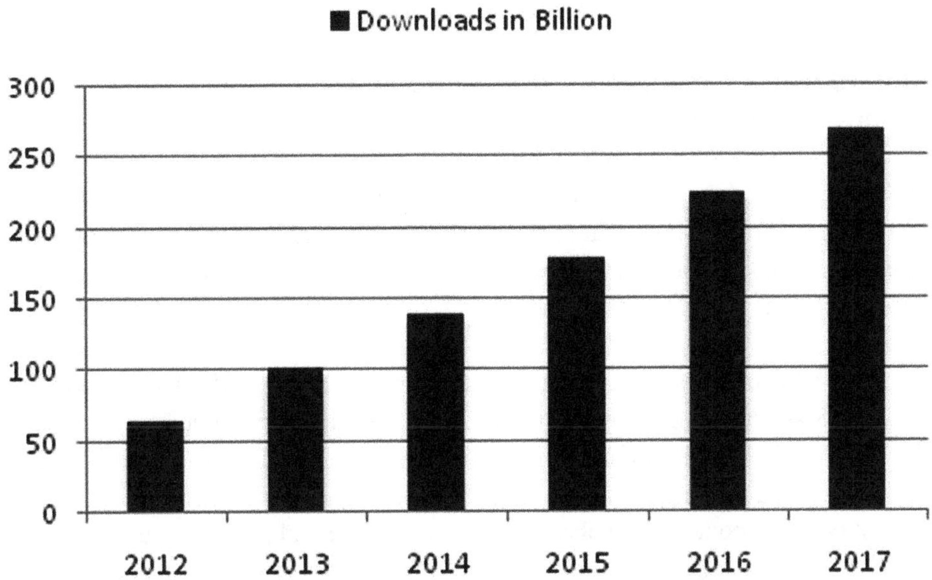

ods. While doing data analysis in biomedical science, software development is an integral part, which involves the development of new software tool to analyze the data, development of databases specific to the dataset, tool development and web server hosting. Software development is done as a collaborative measure between software engineers and biological science researchers which is meant for developing, maintaining and supporting software and data associated with biomedical science. In the process with the help of researchers working in bioinformatics, software engineers analyze the task, understand the nature of information and requirements pertaining to the task and then develop the software or tool.

Software development in bioinformatics has become a big industry and it engage both bioinformatics professionals and computer scientists to encode the power of computational approaches in solving large scale biological problems. Normally JavaScript/AJAX and Java are used for bioinformatics applications, which use software development techniques including test-driven development, perform coding for the test bed experiments, develop tests for all software, create test data/inputs and testing code, implementation and identify the documents requirements and specifications of software in cooperation with other team members. Software is one of the most visible results of bioinformatics research and development. Several recent articles underline the importance of quality code for a real, long-lasting benefit for science. Large quantity of the software has been developed during last ten years which are being used by the bioinformatics professionals for different kind of activities. Hanny et al (2009) and Rother et. al (2012) allude to the fact that between 2006 and 2011 a number of bioinformatics software projects were created.

Mobile App Development on Biomedical Science and Bioinformatics

Development and creation of Mobile applications for bioinformatics has been an active ingredient of biological science research. As alluded to in the introduction this paper looks at some of the mobile apps which are developed specifically for bioinformatics.

- **DNAApp:** Among the numerous applications developed for bioinformatics DNSApp developed by Antibody & Product Development Lab, Singapore is the first mobile application which has been developed to open and analyze DNA sequencing files of ab1. This native app developed for android and iOS can decode and display ab1 sequencing file and provide handy tools like "Reverse Complement", "Chromatogram adjustments", "amino acid translations" and "searching for segments". With the aid of other web based tools this application allows the user to analyze and determine the quality of the sequencing files.
- **MobSBlast:** In order to narrow the gap between desktop applications on sequence alignment and Basic Local Alignment Search Tool or BLAST a mobile app in android was developed with the name MobSBlast. This is very specific to android platform and runs the sequencing algorithm within its application domain and the BLAST services of the query with the sequence databases at the European Bioinformatics institute. MobSBLAST is a bi-modular application where the first module is sequence is similar to functionality of GSA GSA algorithm (Needleman & Wunsch 1970), while the second module of app is based on BasicLoca Alignment Search Tool (BLAST) algorithm (Khanna& Patel 2015). This application is used by life science researchers for finding similarity between molecules.
- **Oh BLAST It!:** This android specific bioinformatics application allows users to work on the taxonomy of a DNA sequence using EBI-EMBL and NCBI's BLAST sequence searching algorithms. When it completes the process it validates the sequence and other search parameters, regularly polls the BLAST sequence search it sent and informs the user when it is completed. (Varambhia, 2013).
- **DNA Alignment:** DNA alignment of simple strings using the Needleman-Wunsch and Smith-Waterman alignment options is done through the DNA

Alignment android application. (Droid 2010). Figure 3 is an illustration of the DNA Alignment in simulator mode.

- **DNA Sequence:** This sequence alignment tool provides the capabilities of local and global alignment in an easy to use interface where users can input two sequences and choose various parameters of alignment for nucleotides or protein sequences (Klein 2014).
- **Genetic Code:** Developed by Damage Inc., is a small application which translates nucleotide condone with a brief overview over the most common amino acids and nucleotides.
- **Genome Voyager:** Genome Voyager is a platform independent web based application that offers an easy access to a collection of whole human genome sequences, allowing researchers to visualize whole genome sequencing data and learn how to analyze them for sequence variants, copy number variations, and loss of heterozygosity events. The access is available through http://www.completegenomics.com/public-data/analysis-tools/genome-voyager/.
- **SNPdbe:** A web based access to SNPdbe database for experimentally and computationally derived annotations on the structural and functional impacts of single amino acid substitutions (Schaefer et al 2012). The mobile and

Figure 3. DNA alignment application in simulator

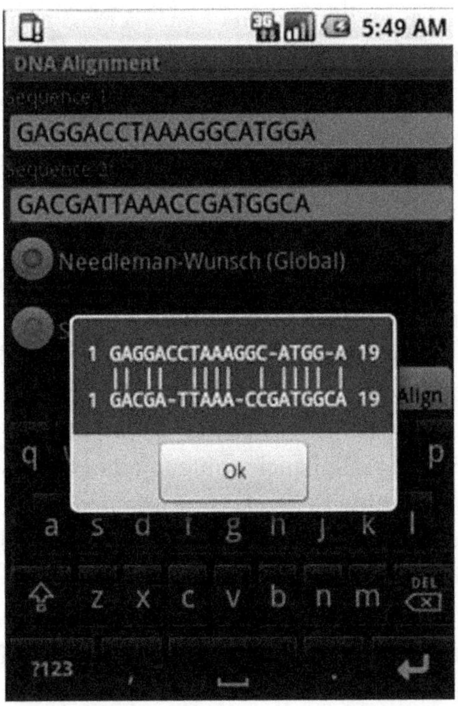

desktop link is available at http://www.rostlab.org/services/snpdbe. This covers over 155000 protein sequences which come from around 2600 organisms where close to a million Single Amino Acid Substitutions (SAASs) are being analyzed for sequencing conflicts.

- **GenomePad: A Mobile UCSC Genome Browser:** GenomePad is and easy-to-use interface to the UCSC Genome Browser. It takes advantage of the important qualities and mobile specific features of iPhone and the genomic maps to perform bioinformatics tasks related to searches in the desired biological species, and maps from genome assembly or specific chromosomal positions. The app can be accessed at http://research.oicr.on.ca/genomepad/.
- **SimGene:** SimGene is an application from Japan Bioinformatics KK is an iOS and Andriod application platform designed for bioinformatics professionals, medical researchers and molecular biologists to do an annotated gene symbol search for over 30 species. It interfaces with Simbiot, Ensembl, NCBI, Gene Ontology, KEGG Pathways, PubMed, Genomic Variations and many other databases to retrieve up-to-date annotated information of search species.
- **RCSB PDB Mobile:** This is a universal app available for both iOS and Android provides fast and mobile access to RCSB Protein Data Bank. It also enables the general public researchers and scholars to search the data bank and visualize protein structures over a wireless or data connection (Quinn et al 2015). Figure 4 is an illustration of RCSB PDB in mobile simulator.
- **FlyExpress App:** This application developed by Arizona and Penn State Universities is designed for both iOS and Android to access images capturing spatial patterns of gene expression during the development of an embryo in the fruit fly (Kumar et al., 2011).
- **iMolview:** This easy to use touch interface for iPhone and Android allows the users to browse Protein Data Bank, DNA and drug molecules in 3D. this has a direct link to DrugBank and Protein Data Bank and allows users to interact with the 3D structures of the biomolecules.
- **Hematopoietic Expression Viewer:** Hematopoietic Expression Viewer allows ready access to the hematopoietic expression data, enhancing the ability to analyze and to identify genes that may be important in hematopoiesis, stem cell biology or blood malignancies among other uses (James et al. 2012).
- **Immunological Genome Project App:** Immunological Genome Project app (Heng, 2008), which includes microarray results but is centered on the immunological mouse community.
- **OnScreen DNA:** This is a suit of 4 applications developed for DNA 3D simulation. It is an interactive 3D simulator for OnScreen DNA Model, OnScreen Gene Transcription and OnScreen DNA Replication.

Figure 4. RCSB PDB Mobile application in simulator

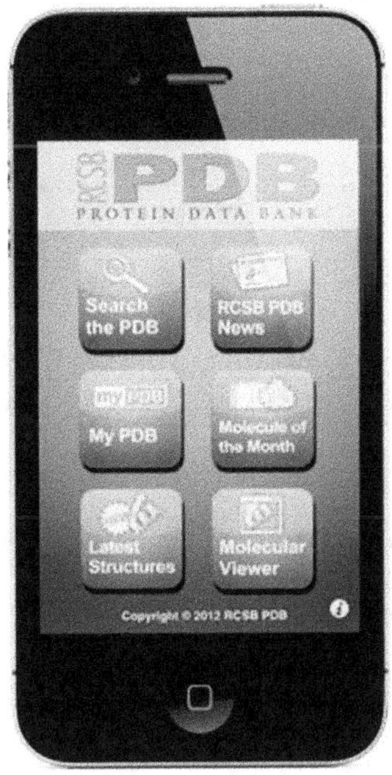

COMMONLY USED

Apart from the preceding specific mobile applications, there are some other mobiles applications which are commonly used by the bioinformatics community. Table 1, Table 2, and Table 3 summarise lists of such applications and utilities.

CONCLUSION

The influence of mobile technology in all genre of the universe of Knowledge is unparalleled. Newer challenges and newer approaches trigger new ideas and the end result is the development of tools and technologies for dynamic data access. Some apps pose different challenges to typical software developers in terms of the efficiency of the app. The advent of big data especially in sequencing methods and in analytics, it is a big question to see whether the computing power of the mobile devices could handle them. Analysis of bioinformatics data depend on very high end sophisticated technologies and specialized computing facilities which makes it a unique field. Can the mobile applications provide such sophistication? This is a question which only time can answer. I am sure with time newer techniques

Table 1. Some of the biomedical apps classified under sequence analysis and alignment

App Name	OS	Description
ALS Online	Android	Align DNA sequences
DNA analyser	Android	Analyses DNA or RNA sequence and GC content
DNA2App - Sequence analyzer	Android	Analyses nucleic acid sequences
DNAApp: DNA sequence analyzer	Android & iOS	Open and analyze DNA sequencing files (ab1)
DNA & Co	Android	Transcript and translate DNA sequences to RNA and protein sequence
DNA Easy	Android	Reverse complement of DNA
DNA Shot	Android	Identify and displays DNA sequences from pictures
DNA to RNA	iOS	Converts DNA sequence into mRNA vice versa
DPSAT	Android	Analyze nucleotide and protein sequences
Gene Aligner	Android	Perform pairwise gene global and local alignments, and generate dot plots for the alignment
Genetic Code	Android	Translate nucleotide codons
SimAlign	Android	Align sequences of genes and proteins
Pairwise Protein Aligner	Android	Generates pair wise protein global and local alignment. A dot plot for the alignment can also be created
Genome	iOS	Sequence instantly and transcribe sample sequences
ETI Bioinformatics	iOS	

Source: Gan, S.K. and Poon, J. 2016.

Table 2. Some of the biomedical apps classified under molecular builder and protein structure viewers

App Name	OS	Description
Atomdroid	Android	Include molecular viewer and builder functions for geometry optimization and Monte Carlo simulation for small molecules
iMolecule Builder	iOS	Support formats from PDB, Sybyl and Crystallographic information. Visualize and build 3D molecules from scratch
iProtein	iOS	Provide access to PDB and Swiss- Prot, RefSeq, Ensembl, etc. Support accurate homology modeling by generating protein structural models
PocketMDS	iOS	Perform molecular dynamics simulation of Lennard-Jones fluids
Yasaraa	Android	Support graphics, molecular modeling and docking
Molecular Dynamics	Android	Perform molecular dynamics simulation of particle motions and thermal changes of the molecular systems
3D-Molecule View	Android	Built on top of jmol librabry and support multiple input formats to visualize 3D molecular structure
Ball&Stick	iOS	Support visualization and local storage of input from PDB
Biochemistry Mnemonics	Android	Provide an interface of the Protein Data Bank to visualize or inspect protein structures in 3D or 2D.

continued on next page

Table 2. Continued

App Name	OS	Description
iMolview Lite	Android & iOS	Browse and view 3D protein and DNA structures
NDKmol - molecular viewer	Android	View three dimensional structures of proteins, nucleic acids and small molecules
PDB Xplorer	Android	Visualize 3D structures of proteins, nucleic acids and small molecules
Pymol	iOS	Display proteins, nucleic acids, and other chemical structures, Support formats including pdb, sdf, mol2, pse, etc.
PDB VIewer		

Source Gan, S.K. and Poon, J. 2016.

Table 3. Examples of biomedical apps classified under database

App Name	OS	Description
ATG Sequence Search	Android	Enables DNA and Protein sequence queries from National Center for Biotechnology Information (NCBI)
Atom 3D	Android	Provide access to the mnemonics database for filtering and editing mnemonics of particular subjects
Harmonizome	Android	Integrate various databases & online resources
iOncology	Android	Provide access to database of enzymatic and cell-based data of several gene families
Mentha the interactome browser	Android	Analysis of selected proteins in the context of a network of interactions.
RCSB PDB Mobileb	Android & iOS	Provide access to RCSB PDB resources
SimGene	Android	Provide up to date, cross reference and integrated genome browser information
PSICQUIC Client	Android	Provide access to the molecular interaction data repository
BioGPS	iOS	Browse gene information
FlyExpressa	iOS	Explore gene expression patterns from Fruit Fly embryogenesis
Yeast Genome	iOS	Browse genes and fundamental chromosomal features of *Saccharomyces cerevisae*
iSpartan	iOS	Provide atomic and molecular properties, NMR and infra spectra, molecular orbitals and electrostatic potential map Model 3D structure of small molecules and estimate energies for alternate conformers

Source Gan, S.K. and Poon, J. 2016.

in mobile technology will be invented to augment the already complicated bioinformatics research. These challenges open new areas for future research, where it would be possible to get answer for some of the burning questions related to the mobile application in bioinformatics and biomedical research will be an interesting observation. With the available resources and the ground breaking work done in the mobile paradigm it is an optimistic sign to look forward to the future developments.

REFERENCES

Barile, L. (2011). Mobile technologies for libraries: A list of mobile applications and resources for development. *College & Research Libraries News*, *72*(4), 222–228.

Droid, B. (2010). *DNA Alignment*. Retrieved from https://play.google.com/store/apps/details?id=blink.dna.align

Gan, S.K., & Poon, J. (2016). The world of biomedical apps: their uses, limitations, and potential. *Scientific Phone Apps and Mobile Devices*, *2*(6).

Gartner. (n.d.). *Top Predictions for IT Organizations 2012 and Beyond*. Retrieved from http://gartner.com/itipage.jsp?id=1328I132010

Hannay, J. E., MacLeod, C., & Singer, J. (2009). How do scientists develop and use scientific software. *ICSE Workshop on Software Engineering for Computational Science and Engineering*, Vancouver, Canada. doi:10.1109/SECSE.2009.5069155

Heng, T. S., Painter, M. W., Elpek, K., Lukacs-Kornek, V., Mauermann, N., Turley, S. J., & Kang, J. et al. (2008). The Immunological Genome Project: Networks of gene expression in immune cells. *Nature Immunology*, *9*(10), 1091–1094. doi:10.1038/ni1008-1091 PMID:18800157

James, R. A., Rao, M. M., Chen, E. S., Goodell, M. A., & Shaw, C. A. (2012). The Hematopoietic Expression Viewer: Expanding mobileapps as a scientific tool. *Bioinformatics (Oxford, England)*, *28*(14), 1941–1942. doi:10.1093/bioinformatics/bts279 PMID:22576171

Jetty, S., & Anbu, K. J. P. (2013). SMS-based content alert system: a case with Bundelkhand University Library, Jhansi. *New Library World, 114*(1-2), 20 – 31.

Khanna, V., & Patel, A. (2015). Mobile Application for Global Sequence Alignmentand BLAST – MobSBlast. *International Journal of Computers and Applications*, *120*(13), 1–5. doi:10.5120/21284-4219

Klein, S. A. (2014). *DNA Sequence*. Retrieved from https://play.google.com/store/apps/details?id=br.com.samuelklein.dna.phonegap.DNA_Sequence

Kroski, E. (n.d.). *On the Move with the Mobile Web: Libraries and Mobile Technologies*. Available online at http://eprints.rclis.org/12463/1/mobile_web_ltr.pdf

Kumar, S., Konikoff, C., Van Emden, B., Busick, C., Davis, K. T., Ji, S., & Newfeld, S. J. et al. (2011). FlyExpress: Visual mining of spatiotemporal patterns for genes and publications in Drosophila embryogenesis. *Bioinformatics (Oxford, England)*, *27*(23), 3319–3320. doi:10.1093/bioinformatics/btr567 PMID:21994220

McQuiggan, S., Kosturko, L., McQuiggan, J., & Sabourin, J. (2015). *Mobile Learning A Handbook for Developers, Educators, and Learners*. Willey Publishers.

Needleman, S. B., & Wunsch, C. D. (1970). A general methodapplicable to the search for similarities in the amino acidsequence of two proteins. *Journal of Molecular Biology*, *48*(3), 443–453. doi:10.1016/0022-2836(70)90057-4 PMID:5420325

Nguyen, P.-V., Verma, C. S., & Gan, S. K.-E. (2014). DNAApp: A mobile application for sequencing data analysis. *Bioinformataics*, *30*(22), 3270–3271. doi:10.1093/bioinformatics/btu525 PMID:25095882

Nicholson, J. A. (2010). The third screen as cultural form in North America. In The Wireless Spectrum: The Politics, Practices, and Poetics of Mobile Media. University of Toronto Press.

Quinn, G. B., Bi, C., Christie, C. H., Pang, K., Prli, A., Nakane, T., & Rose, P. W. et al. (2015). RCSB PDB Mobile: iOS and Android mobile apps to provide data access and visualization to the RCSB Protein Data Bank. *Bioinformatics (Oxford, England)*, *31*(1), 126–127. doi:10.1093/bioinformatics/btu596 PMID:25183487

Ram, S., Anbu, J. P. K., & Kataria, S. (2011). Responding to users expectation in the library: innovative Web 2.0 applications at JUIT Library: A case study. *Program: Electronic Library and Information Systems*, *45*(4), 452–469. doi:10.1108/00330331111182120

Ram, S., Anbu, J. P. K., & Kataria, S. (2014). Mobile Information Literacy for libraries: A case study on requirements for an effective Information Literacy Program. In 5-M Libraries: From Devices to People. Facet Publishing.

Rother, K., Potrzebowski, W., Puton, T., Rother, M., Wywial, E., & Bujnicki, J. M. (2012). A toolbox for developing bioinformatics software. *Briefings in Bioinformatics*, *13*(2), 244–257. doi:10.1093/bib/bbr035 PMID:21803787

Schaefer, C., Meier, A., Rost, B., & Bromberg, Y. (2012). SNPdbe: Constructing an nsSNP functional impacts database. *Bioinformatics (Oxford, England)*, *28*(4), 601–602. doi:10.1093/bioinformatics/btr705 PMID:22210871

Sim, J.Z., Nguyen, P.V., Lim, P.H.J., Su, T.T.C., & Gan, S.K.E. (2015). The Rise of the Mobile Lab: the Use of Smartphone Apps for Biomedical Research. *Asia Pacific Biotech News*, *19*(58).

Varambhia, H. N. (2013). *Oh BLAST It!* Retrieved from https://play.google.com/store/apps/details?id=com.bioinformaticsapp

Zhou, Y., & Ramona, B. (2011). Mobile options for online public access catalogs. *iConference '11 Proceedings of the 2011 iConference*. Retrieved October 8, 2011, from http://dl.acm.org/citation.cfm?id=1940842

Chapter 9
Searching Bioinformatics Information Strategies for Effective Use of Search Engine

Viveka Vardhan Jumpala
Osmania University, India

ABSTRACT

The Internet, which is an information super high way, has practically compressed the world into a cyber colony through various networks and other Internets. The development of the Internet and the emergence of the World Wide Web (WWW) as common vehicle for communication and instantaneous access to search engines and databases. Search Engine is designed to facilitate search for information on the WWW. Search Engines are essentially the tools that help in finding required information on the web quickly in an organized manner. Different search engines do the same job in different ways thus giving different results for the same query. Search Strategies are the new trend on the Web.

DOI: 10.4018/978-1-5225-1871-6.ch009

INTRODUCTION

There is a huge amount of general and biological information on the World Wide Web. The exponential growth of biological data over the past decade has created an enormous challenge to make effective use of the accumulated information. Today bioinformatics is driven by the challenge of integrating the large amount of genetic and structural data emanating from biomedical research. Bioinformatics is the science of storing, retrieving and analyzing large amount of biological information (Buehler, 2005). Bioinformatics refers to the task of organizing, analyzing, and predicting increasingly complex data arising from modern molecular and biochemical techniques. Bioinformatics is a computational analysis of biological information such as nucleic acid and protein sequences and protein structure. Cataloging, classifying, labeling and connecting sequence, structural and functional information of genes and proteins of various organisms will facilitate the discovery of new biological trends. Information search and retrieval is one of the most powerful applications of bioinformatics. The importance of search engines, databases and the increasing sophisticated communication network in biological and biomedical research is tremendous. The ability to use the different online accessible software in molecular biology is becoming mandatory for all biomedical scientists. The current quest to sequence all genes, and to make information available in search engines databases such that all biological investigations must start with browsing the data banks, making computer literacy compulsory for all biologists.

Bioinformatics Definition

According to the Oxford Dictionary website, bioinformatics is conceptualizing biology in terms of molecules (in the sense of Physical Chemistry) and applying "informatics techniques" (derived from disciplines such as applied mathematics, computer science and statistics) to understand and organize the information associated with these molecules, on a large scale. In short, bioinformatics is a management information system for molecular biology and has many practical applications.

Objectives of the Study

The main objective of the chapter is to explore bioinformatics information.

The other objectives of the study are:

1. To find out the various bio informatics databases and search engines.
2. To find out the different types of information search and retrieval strategies of bio informatics information.

3. To find out the bio informatics applications.

Statement of the Problem

An individual cannot read a large amount of data on the web, so users need different types of search strategies to search the bioinformatics information effectively and efficiently.

Significance of the Study

Bioinformatics information becomes an indispensable tool in our everyday life. When we seek bioinformatics information we often go to our favorite search engine or databases and look at the returned pages. This chapter would help to assess the user what type of search strategies use while searching the bioinformatics data to retrieve relevant and exact information from the web.

Methodology

The study is focused on bioinformatics information, and bioinformatics search engines and databases. The study is based on extensive review of literature available in the print journals, online journals on internet to examine about bioinformatics information, bioinformatics databases and search engines, different types of search strategies to search the bioinformatics information from their servers and databases.

Limitations of the Study

Bioinformatics information available globally but the present study is confined to the bio informatics information, different types of bioinformatics search engines and databases and bio informatics applications.

SEARCH ENGINE SEARCH STRATEGIES

The World Wide Web has become an indispensable source of information for any one, it needs to understand how people search and retrieve bioinformatics information. Search engines have been playing an important role in finding the required information from ever growing internet (Henry, 1980).

Bioinformatics search strategies are shown in Table 1.

Table 1. Search engine search strategies

Sl No.	Search Strategies	Description/Use of Search Strategies	Example
1	AND (+ plus sign)	It Narrows search.	Genes AND Drugs
2	OR	Broadens search.	Genes OR Drugs
3	NOT (- sign)	Contain one keyword exclude the other keyword.	Genes NOT Drugs
4	Nesting () Parentheses	Utilizes parentheses to clarify relationships between search terms.	(Genes OR Drugs) AND (Bioinformatics)
5	Proximity Search	Search for two or more words that occur within a specified number of words of each other in the database.	Biomedical Molecular Techniques retrieves records containing three words immediately adjacent to one another and in the same order.
6	NEAR	Find words within 10 words of each other. Near is the same as within 10.	Biomedical near research retrieves records that contain Biomedical and research in any order and within a 10 word radius of one other.
7	BEFORE	Find words in a relative order, specified with the before expression	Biomedical before Research
8	AFTER	Find words in a relative order specified with the after expression.	Biomedical after Drugs
9	Phrase Search	Retrieve search terms next to each other in the order user typed.	"Bioinformatics Research"
10	* Truncation.	Expands a search term to include all forms of a root word.	patent* retrieves patent, patents, patentable, patented, etc.
11	Multi Charcter Wild Card *	Multi-character wildcard for finding alternative spellings.	behavi*r retrieves behaviour or behavior
12	Stop words	Stop words are ignored	a, and, the
13	File Format Search	Users can limit their search to any specific file format.	MicrosoftWord (.doc), Adobe Pdf (.pdf), Microsoft Excel (.xls), Text Format(.txt) etc.
14	Maps	related maps can be displayed	Bioinformatics:Map
15	Language	Search can be limited by language.	English, Hindi
16	Spelling Check	Mistake in spelling then system asks 'did you mean this'.	Bioinformtcs Did you mean Bioinformatics
17	Images	Relevant images	Drugs
18	Dictionary Lookup	"define" followed by a colon and the word(s) to look up	Define: Bio-informatics
19	Google Goggles	Google app	Google app photos
20	Search web pages with a specific domain extension	Search by domain with in education sector websites (.edu), or Government (.gov), or information (.info), or commercial (.com) etc.	(.edu) (.gov) (.info) (.com)

BIOLOGICAL DATABASE ORGANIZATION

The exponential growth of biological data over the past decade has created an enormous challenge to make effective use of the accumulated information. Information stored must be correct, complete and internal relationships among elements easy to navigate. Computational tools, search engines, and databases are essential to the management and identification of patterns among database elements that reflect biological system (Gautham, 2007). The National Center for Biotechnology Information (NCBI) in the United States and the European Bioinformatics Institute (EBI) in England are two main life science servers responsible for dealing with the staggering volume of data. There are various bioinformatics search engines and databases shown in Table 2, as well as multiple bioinformatics projects as seen in Table 3.

The currently available data for DNA and protein sequences is so enormous that searching for information is dubbed, "biological data mining" similar to a real gold mine with hidden treasures where common rocks have to be separated from gold nuggets. Search engines perform two basic tasks:

1. Simple string searches for information retrieval of stored data [GenBank (nucleotides and proteins), PubMed (MEDLINE), 3-D structures, genomes, and taxonomy databases].
2. Smilarity searches (e.g. Blast) to retrieve, align and compare sequences or structures.

Applications for Bioinformatics

- Knowledge based drug design,
- Forensic DNA analysis,
- Agricultural biotechnology, and
- Computational studies of protein ligand interactions.

CONCLUSION

Bioinformatics have undergone astonishing development in the recent years is enormous. Many computational algorithms, methods and search strategies have been adopted for biological research. Bioinformaticians, researchers need to know different search strategies for effective use bioinformatics search engines (Grossman, 2009). Bioinformatics in the field of medical research, two major challenges, first the analysis of new data delivered by high-throughput technologies and second the combination and integration of different data types to achieve a more holistic

Table 2. Bioinformatics search engines/databases

Sl. No.	Name of the Search Engine/Database	Description/Usefulness	Web Address/URL
1	Bioinformatics Harvester	Harvester collects information from protein and gene database from prediction servers.	http://bioinformatics.ca/links_directory/tool/9872/harvester
2	BioText Search Engine	scientific literature searches more than 300 open access journals.	http://biosearch.berkeley.edu/
3	BioPortal	Web accessible open repository of biomedical ontologies	http://bioportal.bioontology.org/
4	BioCyc	BioCyc is a collection of 7667 Pathway/ Genome Databases (PGDBs).	http://biocyc.org/
5	ChemBank	ChemBank is a web-based informatics for chemical genetics	http://chembank.broadinstitute.org/
6	ChEBI	Chemical Entities of Biological Interest (*ChEBI*) is a freely available dictionary of molecular entities focused on 'small' chemical compounds.	https://www.ebi.ac.uk/chebi/
7	COMPEL	Composite regulatory elements in eukaryotes is a database on composite regulatory elements providing combinatorial transcriptional regulation.	http://www.ncbi.nlm.nih.gov/pubmed/
8	DOGS	Database of genome sizes	http://www.cbs.dtu.dk/databases/DOGS/
9	eTBLAST	It is a textual similarity search engine.	http://www.ncbi.nlm.nih.gov/pmc/articles
10	ENTREZ	Integrated search engine of National Centre for Biotechnology Information	www.ncbi.nlm.nih.gov/Entrez
11	EMBL-EBI	Provides data from life sciences, performs basic research in computational biology.	http://www.ebi.ac.uk/
12	EPD	Eukaryotic promoter database	http://www.edp.isb-sib.ch/
13	GOLD	Information on genome projects around the world	http://wit.integratedgenomics.com/GOLD/
14	GDB	The Genome database	http://gdbwww.gdb.org/
15	IMAGE	The largest collection of DNA sequence clones	http:image.Ilnl.gov/
16	IMGT	The international Immune Genetics Information system	http://imgt.cines.fr/
17	KEGG	Kyoto Encyclopedia of Genes and *Genomes* is a bioinformatics resource for linking *genomes* to life and the environment.	http://www.genome.jp/kegg/
18	Locuslink	Single query interface to sequence and genetic loci	http://www.ncbi.nlm.nih.gov/LocusLink/
19	MitoDat	Mitochondrial nuclear genes	http://www-Immb.ncifcrf.gov/mitoDat/

continued on next page

Table 2. Continued

Sl. No.	Name of the Search Engine/Database	Description/Usefulness	Web Address/URL
20	Medical Subject Heading (MeSH)	(MeSH) is a comprehensive controlled vocabulary for indexing journal articles and books in the life sciences.	https://www.nlm.nih.gov/mesh
21	NDB	Nucleic acid database	http://ndbserver.rutgers.edu/
22	PDB	Protein Data Bank is an information portal to 115918 Biological Macromolecular structures of proteins, nucleic acids, and complex assemblies.	http://www.rcsb.org/pdb/home/home.do
23	Refseq	The NCBI reference sequence project	http://www.ncbi.nlm.hih.gov/LocusLink/refseq.html
24	SMART	Simple Modular Architecture Research Tool	http://smart.embl-heidelberg.de/
25	SHIGAN	Shared Information of Genetic resources, Japan	http://www.grs.nig.ac.jp/
26	TOXNET	(TOXicology Data NETwork) is a group of databases covering chemicals and drugs, diseases, environmental health, occupational safety, poisoning, and toxicology.	https://www.nlm.nih.gov/pubs/factsheets/toxnetfs.html
27	TIGR	Curated databases of microbes, plants and humans	http://www.tigr.org/tdb/index.html
28	TRRD	Transcription Regulatory region database	http://wwwmgs.bionet.nsc.ru/mgs/dbases/trrd/4
29	Transfac	The transcription factor database	http://tranfac.gdf.de/TRANSFAC/index.html
30	Uniport	Universal Protein Resource Largest Protein Database	http://www.uniprot.org/
31	Unigene	Cluster of sequences for unique at NCBI	http://www.ncbi.nlm.nih.gov/genome/sts/
32	VADLO	Life Sciences Search Engine. It searches the bio informatics databases. Search for methods, techniques and protocols.	http://vadlo.com/

Table 3. Bioinformatics projects

1	The Whole Brain Atlas	www.med.harvard.edu/AANLIB
2	The Human Brain Project	www.gg.caltech.edu/hbp
3	International HapMap Project	www.hapmap.org
4	Tree of Life Project	http://tolweb.org/tree
5	Gene Ontology Consortium	www.geneontology.org
6	Pharmacogenetics Research Network	www.pharmGkb.org

picture of a disease. The integration of information science and molecular biology will intensify as faster computers and internet connections facilitate biological research (Vittal, 2005). The challenges are the construction of a theoretical basis of cellular interactions and the integration of various types of data.

FUTURE DIRECTIONS

Bioinformatics future directions are storage and analysis of biological data will continue to a important task for bioinformatics. Computational systems in bioinformatics will go far beyond their present scope and include extended models of biological systems. Bioinformatics scientists, researchers need to discover the new hidden knowledge in the miraculous world of molecules, particularly in the field of medical research patient information to predict treatment outcome for future generations.

ACKNOWLEDGMENT

The author would like to thank professors Dr. Chandrashekar Rao, Dr. N. Laxman Rao, Dr. Sudarshan Rao, and Dr. V. Vishwa Mohan, for their able guidance, encouragement, constant support and whole-hearted cooperation.

REFERENCES

Buehler, L., & Rashidi, H. (2005). Bioinformatics basics: Applications in biological science and medicine. Taylor and Francis Group.

Center for Biotechnology Information. (n.d.). Retrieved from https://en.wikipedia.org/wiki/National

Gautham, N. (2007). *Bioinformatics Databases and Algorithms*. Delhi: Narosa Publishing House Limited.

Grossman, D. A. (2009). Information Retrieval: Algorithms and Heuristics (2nd ed.). Springer International Edition.

Henry, W. M.. et.al (1980), *Online searching: An introduction*. Butterworths and Company.

Vittal, R. S. (2005). Bioinformatics a modern approach. Prentice Hall of India.

Chapter 10
Information Seeking Behavior of Medical Scientists at Jawaharlal Nehru Institute of Medical Science:
A Study

Bobby Phuritsabam
Manipur University, India

Arambam Bidyaluxmi Devi
Manipur University, India

ABSTRACT

Purpose: The purpose of the study is to identify the library services and facilities provided to the Medical Scientists of JNIMS, Porompat. The study is limited to Medical Scientist of JNIMS who employed at twenty two (22) different medical departments of JNIMS. Design/Methodology/Approach: The study is based on survey method; questionnaire and interview method is used for collection of primary data. Hundred (100) questionnaires were distributed to the medical scientist of JNIMS. Findings: Services and facilities provided by the library are not satisfied by the medical scientist; library lack qualified manpower to function the library. Originality/Value: The study is part of the dissertation submitted to the Department of Library and Information Science, Manipur University for the year2014-2015. Article Type: Case Study

DOI: 10.4018/978-1-5225-1871-6.ch010

INTRODUCTION

Information is a valuable resource of today's Information Society, thus acquiring, using and implementing information are critical activities. The process is known as information seeking; therefore it is becoming more fundamental and strategic for intellectual activities. Information seeking process is very much dependent on the interaction among information seekers and other professionals and system representing information. The term information has been derived from the Latin words "Forma & Farmatio". Information means the communication of knowledge about an event or given condition or the spread of knowledge derived from observation, study or instruction.

Information seeking behavior is a broad term encompassing the way individuals articulate their information need, seek, evaluate, select and use information. The information-seeking behaviour by human being has been elaborated and defined differently by many eminent authors and scholars from titme to time. According to Marchionini and Komlodi (2001, p. 25) "information seeking is a process in which humans engage to purposely change their state of knowledge." The process is inherently interactive where the information seekers is impacted by the direct attention, accept and adapt to stimuli, reflect on progress and evaluate the efficacy of continuation. This definition was closely suppoted by many scholars and researcher such as Case (2002), Kari and Savolainen (2003), Kuhlthau (1991), Sujatha (2014) and Vakkari (1998). This definition is the one most closely aligned and discussed by Foster (2004) as non-linear model of information seeking and stated as "Information seeking was found to be framed by the resolution of the information problems, and by limits to time and financial resources".

Kumar (2004) Information seeking behavior results from the recognition of some need, perceived by the user. Kumar, Nasima & Sukhleen (2004) as any activity of an individual that is undertaken it identifies and perceives that the current state of processed knowledge is less than that needed to deal with some issue. Kumar (2004) emphasized that the information seeking behavior is mainly concerned with who needs what kind of information and for what reasons; how information is found, evaluated and used, and how their needs can be identified and satisfied. Medical as a professional is one of the most dynamic fields where new medicines and investigation keep coming daily. Medical professional facing the major problem is that they had less time for self-study and needed a system to which they can be updated without disturbing their routine work. The study aims to study the needs and seeking behavior of the Medical professionals.

LITERATURE REVIEW

There have been extensive and enormous studies have been published about the information seeking behavior of students in libraries. Some attempts have been made to cover studies on area of medical sciences too. These studies pertain to the discussions of the information seeking behavior, and their direct applications to the awareness and utilization of the print or digital resources. User training is quite helpful in obtaining useful, and relevant information is an outcome of the study reported by Asemi (2005) at MUI in Iran. Dong (2003) studied the use of internet resources and the evaluation of its usefulness by Chinese students' and academics'. Curtis et al. (1997) found that health sciences faculty's information seeking behavior depends on upon the currency of information they needed including the use of new information technologies at the University of Illinois in Chicago. The impact of the Internet on information seeking behavior of medical students and faculty and their medical library use was studied by Tao et al. (2003).

Joanee (2005) studied information seeking behavior of biology graduate students. The study shows that there are differences in the extent to sources of information were used by students in different years of their studies. Singh (2007) while conducting a study on agricultural science shows that agriculture scientist has expressed great dependence in meeting their information requirements in their institutional library/ information center. Mojtaba & Sookhtalo (2009) studies reveals the awareness of scientific resources and availability of library resources items was the most important. Maynard (2012), presented a report of his research conducted to model the information seeking behavior of graduate students of Kuwait University and the factor influencing the behavior. Islam (2012), investigated the information seeking behavior of print media journalist. The study attempts to identify how successful the journalists were and what information source was preferred. The study suggested that journalism schools should include courses in their curriculum about information behavior various types of information sources, information retrieval, and search strategies. Maharana (2013), the main objectives of the study were to examine the frequency of visit to the library of medical practitioners and services offered by the libraries. The study identifies that awareness of resource, ability to use the tool, self-evaluation, were a positive impact on Medical Practitioners information needs and seeking. Kumar (2014) reviewed the studies undertaken to identify the information seeking behavior of the research scholars and faculty members of Kurukshetra University. The study suggests that the library should organize a training program for information professionals, the speed of internet should be increased and provide a sufficient number of computer to improve the Information seeking behavior. With this broad literature review, it was assumed that there must be study to identify the information needs of the users working in the area of medical sciences.

With the advent of technology, there has been paradigm shift in the information seeking behavior. The internet has revolutionized the way people looks for information. At the same time advancement of mobile internet and mobile technology has act as kindle fire into the information seeking behavior. Mobile technology has advanced the the people look for information, specially health related information. Number of studies have shown that the people use online health information very frequently and share what they have learned and experienced through mobile devices (Hendrick, 2011). Due to deep penetration of mobile phones into peoples daily life, the the mobile phones are conveniently being used for consumers to seek health information (Whittaker and Smith, 2008; Cocosila and Archer, 2010). There are some other incidence reported where researchers have repoted their results pertaining to the health-related issues being discussed over the internet (Wilson and Risk, 2002; Liang et al., 2011) and using mobile phones (Xue et al., 2012). Consumers tend to search for health-related web sites and information that were previously unavailable. Such as advancement of technology applications in health related issue motivated further to undertake such a study.

The Jawaharlal Nehru Institute of Medical Sciences (JNIMS), is located at Imphal-East District, Manipur. The government of India has given the Letter of Permission (LOP) through the Medical Council of India (MCI) vided. MCI No. 34(41)/2010-Med.-907 Dated 14th July 2010. Manipur University, Canchipur given permission to affiliated JNIMS for the year 2013-2014. At present JNIMS have twenty-four (24) different departments.

JNIMS is managed by the Jawaharlal Nehru Institute of Medical Sciences Society, under the Government of Manipur. The Society was registered on 18th December 2007. The General Body of the society, known as Governing Council of the Society, is presided by the Hon'ble Chief Minister of Manipur, with Director, NIMS as the member Secretary. The institute library is well stocked with books, monographs, and journals with 25 nodes of internet connections. The library is fully air-conditioned. There is the provision of –journals and skill lab. There are two (2) reading room: one internal and one external with a capacity of hundred (100) students each. It has a collection of more than 38,424 books, Ninety Six (96) journals, and four hundred seventy-seven (477) CD-ROM.

OBJECTIVE OF THE STUDY

The study aimed to identify the following:

1. The needs of information and seeking behavior of the Medical Scientists of JNIMS, Manipur, with specific objectives to know the preferred source of information used by the Medical Scientists of JNIMS.
2. To study the behavior of information seeking.
3. To study the facilities required by the Medical Scientists.
4. To assess the level of satisfaction of library services.

METHODOLOGY

In order to meet the objectivs, the study was conducted through survey method collect data. The primary data collected through the open-ended questionnaire. The questionnaires were primarily divided into four sections:

1. Personal Information,
2. Information Seeking Behavior,
3. Source and services, and
4. Problem and suggestion.

DATA ANALYSIS AND INTERPRETATION

The data analysis is based on the sample collected from different medical scientists from JNMC. A total of 100 semi-structured questionnaire was distributed, out of this 59 duly filled responses were received and out of these five were incomplete and excluded from data analysis. Out of these fifty four responses, 28(51.85%) were male and 26 (48.14%) were female.

THE FREQUENCY OF VISITING THE LIBRARY

Visiting frequency is one of the most important criteria for assessing the use of the library and resources. It was tried to get the answer of the frequency of library visits, it was found that 59% of the respondent go to the library occasionally, 25% go to the library weekly, 9% respondent visit the library thrice in a week, and 5% visit the library daily.

THE PURPOSE OF LIBRARY USE

Library is a center for resource and research. In general perception, a user visits library either reading or borrowing books. When asked about the purpose of the visit of the library, 70.37% respondent replied that they visit the library for updating their knowledge; 44.44% visits for preparing class lecture with the help of library resources and 20.37% visits the library for access resources for their research work. There were 16.66% users who visit library for looking information on treatment practices and 12.96% visits for guiding researchers. The majority of the respondent are searching information about current changes and new development in their field, identifying and treatment of new diseases. 88.88% are using the information at the departmental library, and 33.33% are using information at home; 5.55% at the cyber café and 3.70% at the State information library.

Students' Awareness of Digital Resources in General

Students were asked whether they were aware of digital resources in general. It was found that 70% respondents are aware of digital resources. A total of 15 users (30% percent) gave a negative response.

METHODS AND INFORMATION SOURCES USED FOR CURRENT INFORMATION

Library has a huge collection of books, journals, and research materials. 88.88% access the information by reading latest book; 79% access information by consulting experts in the subject field; 85% used the information sources such as attending conference, seminar, workshop and 79% of the respondent access the information by using library resources.

- **Use Pattern of Various Documents:** The students were asked how they used the various document, and it is found that 59% of the respondent uses institutional resources, 53% use online e-resources while 27% use documentary resources. According to research by Groote and Dorsch (2003), the success of the online resources depends upon the databases subscription having bibliographic information.
- **Preferred Format of Information:** Most of the user responded that they prefer print resources (79%) as compared to Digital Resources (21%). There has always been a big question mark on the usage of electronic resources such

as e-journals and ebooks. Various studies have highlighted this concern on the low usage of electronic format (Wu & Chen 2011; Nicolas et al. 2008).

- **Preferred Tool for Searching Information:** There is any number of bibliographic databases and online public access catalogs (OPACs) available on the internet. Most libraries use commercial document delivery services to ensure quick and efficient access to primary information, especially for the researchers. It is found that 62.96% used indexing and abstracting journals to access information, 57% uses medical website, 50% uses references to book and journals, 46% used library catalog and 29% use bibliographies to find out relevant information.
- **Problems Faced by Users in Information Seeking:** The students were asked to indicate problem faced for information access, 55.55% of the responded indicate the required information/materials are not available at the library, 24.07% reported the inability of the information search as the information is scattered in too many sources available at the library,38.88% indicate that the latest information sources are not available at the library, 25.92% indicate lack of time in searching the information, 27.77% encountered irregular power supply, 3.70% indicate lack of library staff non-cooperation in searching information and 3% indicate lack of training & awareness in e-resources available at the library.

LEVEL OF SATISFACTION IN ACCESSING LIBRARY INFORMATION BY THE MEDICAL SCIENTISTS

Table 1 shows that the majority of the satisfactions level of accessing the information by the Medical Scientist of JNIMS are partially satisfied with the services and facilities provided to the Medical Scientist.

CONCLUSION

The authors surveyed a sample of 54 students of JNMC Manipur University ascertain their awareness of medical information. It was found that 70.37% of the respondents access information for updating knowledge; most of the Medical Scientists visit library occasionally; 55.55% of the respondents are searching information about the current changes and new development; 66.66% of the respondent are always using internet services daily; 90.74% are using departmental library; 55% indicate that the required materials are not available at the library. The library should pay attention to the latest and current information; Trained and qualified staff must

Table 1. Level of satisfactions (N=54)

Facilities	Fully Satisfactory		Partially Satisfactory		Not Satisfactory	
	Respondent	Percentage	Respondent	Percentage	Respondent	Percentage
Institutional Library	13	24%	33	61%	8	14.81%
Departmental Library	15	27%	34	62%	5	9.25%
Record Unit/ section	3	5.55%	29	53.70%	22	40.74%
Information Unit	1	1.85%	20	37%	33	61%
Own Personal Collection	12	22.22%	14	25.92%	9	16.66%
Internet Connectivity	8	14.81%	18	33.33%	28	51.85%

employ to cater the needs of the Medical scientists. The library should organize training program & user education program to create awareness about the digital resources. The power supply in the library should be proper to create an ambiance of learning along with strong internet services. If such steps are taken, it will help the library to attract new users.

REFERENCES

Anil Kumar, N.S. (2014). Information seeking behavior by the research scholars & faculty members: A survey of Kurukshetra University, Kurukshetra in the disciplines of life science. *IOSR Journal of Humanities and Social Sciences, 19*(6), 119-138.

Asemi, A. (2005), Information searching habits of Internet users: a case study on the Medical Sciences University of Isfahan, Iran. *Webology*, *2*(1).

Callinan, J. E. (2005). Information Seeking Behavior of Undergraduate Biology Students. *Library Review*, *54*(2), 86–99. doi:10.1108/00242530510583039

Case, D. O. (2002). *Looking for Information: A Survey of Research on Information Seeking, Needs, and Behavior.* Amsterdam: Academic Press.

Cocosila, M., & Archer, N. (2010). Adoption of mobile ICT for health promotion: An empirical investigation. *Electronic Markets*, *20*(3/4), 241–250. doi:10.1007/s12525-010-0042-y

Curtis, K. L., Weller, A. C., & Hurd, J. M. (1997). the Information-seeking behavior of health sciences faculty: The impact of new information technologies. *Bulletin of the Medical Library Association*, *85*(4), 402–410. PMID:9431430

Dorsch, J. L. (2000). Information needs of rural health professionals: A review of the literature. *Bulletin of the Medical Library Association*, *88*(4), 346–354. PMID:11055302

Foster, A. E. (2004). A nonlinear model of information seeking behaviour. *Journal of the American Society for Information Science and Technology*, *55*(3), 228–237. doi:10.1002/asi.10359

Hendrick, B. (2011). *Internet popular with people seeking health information*. Available at: http://women.webmd.com/news/20110512/Internet-popular-with-people-seeking-health-information

Islam, M. A. (2012). Information-seeking by print media journalist in Rajshahi, Bangladesh. *International Federation of Library Association and Institution*, *38*(4), 283–288.

Kari, J., & Savolainen, R. (2003). Towards a contextual model of information seeking on the web. *The New Review of Information Behaviour Research*, *4*(1), 155–175. doi:10.1080/14716310310001631507

Kothari, C. R. (2004). Research Methodology Methods and Techniques (2nd ed.). New Delhi: New Age International Publishers.

Kuhlthau, C. C. (1991). Inside the search process: Information seeking from the users perspective. *Journal of the American Society for Information Science*, *42*(5), 361–371. doi:10.1002/(SICI)1097-4571(199106)42:5<361::AID-ASI6>3.0.CO;2-#

Liang, H., Xue, Y., & Chase, S. K. (2011). Online health information seeking by people with physical disabilities due to neurological conditions. *International Journal of Medical Informatics*, *80*(11), 745–753. doi:10.1016/j.ijmedinf.2011.08.003 PMID:21917511

Marchionini, G., & Komlodi, A. (2001). Design of interfaces for information seeking. *Annual Review of Information Science & Technology*, 33.

Maynard, N. A. M. (2012). Modeling information-seeking behavior of graduate students at Kuwait. *The Journal of Documentation*, *68*(4), 430–459. doi:10.1108/00220411211239057

Mojtaba Sookhtanlo, H. M. (2009). Library information–seeking behavior among Undergraduates students of Agricultural extension and education in Iran. *DESIDOC Journal of Library & Information Technology*, *29*(4), 12–20. doi:10.14429/djlit.29.256

Nel, J. T. (2001). The information seeking process: Is there a sixth sense? *Mousaion*, *19*(2), 23–32.

Nicholas, D., Rowlands, I., Clark, D., Huntington, P., Jamali, H. R., & Ollé, C. (2008). The UK scholarly e-book usage: A landmark survey. *Aslib Proceedings*, *60*(4), 311–334. doi:10.1108/00012530810887962

Rabindra, K., & Maharana, A. P. (2013). Exploring the information seeking behavior of Medical Science. *Information Age*, *7*(1), 18–22.

Singh, K. S. (2007). Information seeking behavior of Agricultural scientists with particular references to their information seeking strategies. *Annals of Library and Information Studies*, *54*, 213–220.

Sujatha, S. (2014). Exploring information seeking behaviour in the changing ICT environment: A snapshot of Kakatiya University facility. *International Journal of Information Services and Technology*, *1*(1), 11–14.

Tao, D., Demiris, G., Graves, R. S., & Sievert, M. (2003). Transition from in library use of resources to outside library use: the impact of the Internet on information seeking behavior of medical students and faculty.*AMIA Annual Symposium Proceedings*.

Thanuskodi, S. (2012). Use of Internet and Electronic Resources among Medical Professionals with special reference to Tamil Nadu: A Case Study. *SRELS Journal of Information Management*, *49*(3), 281–292.

Vakkari, P. (1998), Task complexity, information types, search strategies and relevance: integrating studies on information seeking and retrieval. In *Exploring the Contexts of Information Behaviour,Proceedings of the Second International Conference on Research in Information Needs, Seeking and Use in Different Contexts*.

Whittaker, R., & Smith, M. (2008). M-health – using mobile phones for healthy behavior change. *International Journal of Mobile Marketing*, *3*(2), 80–85.

Wilson, P., & Risk, A. (2002). How to find the good and avoid the bad or ugly: A short guide to tools for rating quality of health information on the internet. *British Medical Journal*, *324*(7337), 598–602. doi:10.1136/bmj.324.7337.598 PMID:11884329

Wu, M., & Chen, S. (2011). Graduate students usage of and attitudes towards e-books: Experiences from Taiwan. *Program*, *45*(3), 294–307. doi:10.1108/00330331111151601

Xue, L., Yen, C. C., Chang, L., Chan, H. C., Tai, B. C., Tan, S. B., & Choolani, M. et al. (2012). An exploratory study of ageing womens perception on access to health informatics via a mobile phone-based intervention. *International Journal of Medical Informatics*, *81*(9), 637–648. doi:10.1016/j.ijmedinf.2012.04.008 PMID:22658778

Yogesh, K. S. (2006). *Fundamental of Research Methodology and Statistics*. New Delhi: New Age International Publishers.

KEY TERMS AND DEFINITIONS

E-Journals: The journals available in electronic format.

Information Literacy: Mechanism adopted to inform users about resources, services, and activities of the library.

Medical Library: The library system dedicated to medical sciences.

Chapter 11
Information Needs and Assessment of Bioinformatics Students at the University of Swaziland:
Librarian View

Satyabati Devi Sorokhaibam
University of Swaziland, Swaziland

Ntombikayise Nomsa Mathabela
University of Swaziland, Swaziland

ABSTRACT

A survey was carried out of the information landscape within the students of Computer Science, Biology and Mathematics in the University of Swaziland which examined the research problems, important sources of information, the methods of access, information needs and seeking behavior of the users their assessment and the role of the Libraries since Librarian have to identify the information needs, uses and problems faced to meet the needs and requirement of the user. A total of 200 questionnaire were distributed. The survey indicated that majority of the students believe that the online resources play a very important role for their research and show positive attitude toward future bioinformatics usage and training. The study concluded that the training preferences of students need to be further explored.

DOI: 10.4018/978-1-5225-1871-6.ch011

INTRODUCTION

Information plays a vital role to achieve the existing information technologies. As bioinformatics has become an important discipline in biological sciences several developing countries have been making progress in this field lately. The librarians have to identify the needs of the user and the problems encounter to meet the requirement of the user in the present age of information. Information needs and uses should be the focus of attention of the librarian to meet the specific needs of the individual or a specific group.

For a proper and systematic planning and development of information resources and services user studies are quite important to know the basic needs of the users and it is also important for the librarian to keep track in all the emerging discipline. The present study is focuses on the information needs and assessment of bioinformatics users of the University of Swaziland to design an information system and also in building up need based information for the students.

The study follows questionnaire methods which consist of 15 questionnaires with their personal details and focus mainly on needs, uses and identification of various problems arising out of it.

OBJECTIVES OF THE STUDY

- To investigate the methods and sources used by students to acquire required information.
- To find out the importance of various information resources for their academic and research activities.
- To study their information gathering activities.
- To what extend does the students used the current range of online database resources that is available to them.
- To find out their familiarity with the attributes of Printed and Electronic Information Sources.
- To ascertain what problems are encountered by the students in seeking information.

SCOPE

The scope of the study is limited to the students of Computer science, Engineering, Biology and Mathematics. A total of 200 questionnaires are distributed to the students randomly. Out of which 152 (76%) are received and used for the analysis.

METHODS

This study was exploratory in nature therefore both quantitative and qualitative methodological tools were employed. A questionnaire survey was conducted which was followed by interview in spite of the users heavy and tight schedule. The questionnaire was intended to elicit the nature and type of information that bioinformatics users' need and uses in order to carry out their research and based on the objectives of the study. It also sought to ascertain what information resources and services they would find useful to accomplish their information needs. The collected data is arranged, analyzed and interpreted by employing statistical methods to draw inferences that were formulated and also to fulfill the stated objectives of the study.

DATA ANALYSIS AND INTERPRETATION

The demographic profile of the respondents includes 90 (45%) were male and 62(31%) were female. For the question about the awareness of the electronic resources, most of the student were aware of the term (Table 1).

From the above analysis, it is seen that 90% male and 92% female were aware of the e-resources available in the University. Only 8% male and 5% female were not aware. Awareness of e-resources depends upon the marketing strategies adopted by the library, still there are other factors also responsible for non-awareness. This lack of awareness may be because of awareness of the general resources results from Internet access being expected to be available at the library from a theoretical point view (Renwick, 2005). Further, the result of non-awareness may be the reason, where students may not present during the library skill classes.

The students were asked whether they are satisfied with the service available in the library. More than half of the student's were not satisfied with the library services (Table 2). The reason could be assumed that the lack of awareness about the various sercices. Most of them complain about the collections.

The Table 2 indicates that only 38% male and 37% female were satisfied with the services while 54% male and 53% female were not satisfied. It shows that the library needs to market their services more.

Table 1. Awareness of e-resources

Sl. No.	Yes		No		No response	
	Male	Female	Male	Female	Male	Female
1.	81 (90%)	57 (92%)	7 (8%)	3 (5%)	2 (2%)	2 (3%)

Table 2. Satisfaction about the library services

Yes		No		No response	
Male	Female	Male	Female	Male	Female
34 (38%)	23 (37%)	49 (54%)	33 (53%)	7 (8%)	6 (10%)

Table 3 shows the frequency of visit to the library. It was observed that majority of the students 41% male and 37% female visited 2-3 times a week while 28% male and 26% female visited weekly. It is not encouraging to see the frequency of visit. The frequency of library visit and resources access often visualized as a measure of student success (Cox & Jantti 2012; Goodall & Pattern 2011; Stone, Ramsden and Pattern, 2011).

From Table 4, it is clear that the purpose of visit to the library is mainly for self-study and to borrow books. At least 37% male come to browse the resources while

Table 3. Frequency of visit

Frequency	No. of Respondent	
	Male	Female
Everyday	14 (16%)	8 (13%)
2-3 times	37 (41%)	23 (37%)
Weekly	25 (28%)	16 (26%)
Monthly	11 (12%)	12 (19%)
No response	3 (3%)	3 (5%)

Table 4. Purpose of visit (tick as many as applies to you)

Purpose	No. of Respondent	
	Male	Female
Self Study	69 (77%)	48 (77%)
Borrowing Books	66 (73%)	44 (71%)
Browsing the resources	33 (37%)	10 (16%)
Use of reference materials	14 (16%)	16 (26%)
Read journals	11 (12%)	11 (18%)
Collect course materials	7 (8%)	8 (13%)
Read newspaper and magazines	28 (31%)	12 (19%)
No response	1 (1%)	2 (3%)

the female stands for 16% only. It was surprising to note the percentage of students come to read the journals is 12% male and 18% female.

Table 5 shows the use of library sources and services. Majority of the students use the sources books/ journals (59% each male & female). Issue/return and references searching services have been used by more female (31%) than male (17%).

Table 6 shows the purpose of seeking information. It indicates that majority of the students 66% male and 73% female seeks information to write an article. While 46% male and 44% female come to solve immediate problems. The immediate problems were mainly the class assignments for their class assessment.

Table 7 depicts the preference of journals. Most of the students prefer the e-journals from the print journals. 63% male and 73% female want the e-journals while still 30% male and 23% female prefer the print one. This shows that the print

Table 5. Use of library source and service

Services	No. of Respondent	
	Male	Female
Issue / Return	15 (17%)	19 (31%)
Reference Service	16 (18%)	16 (26%)
Books / journals	53 (59%)	37 (59%)
Seminar/ Conferences	0 (0%)	1 (2%)
Indexing Abstracting	1 (1%)	0 (0%)
Bibliography	7 (8%)	4 (7%)
Interlibrary loan	1 (1%)	2 (3%)
No response	7 (8%)	7 (11%)

Table 6. Purpose of seeking information

Purpose	No. of Respondent	
	Male	Female
For career development	19 (21%)	13 (21%)
To solve immediate problem	41 (46%)	27 (44%)
To keep up to date	21 (23%)	8 (13%)
To write an article	59 (66%)	45 (73%)
No response	3 (3%)	3 (5%)

Table 7. Preference of journals

E-Journals		Print		No Response	
Male	Female	Male	Female	Male	Female
57 (63%)	45 (73%)	27 (30%)	14 (23%)	6 (7%)	3 (5%)

cannot be stop until and unless the user are get use to the e-journals. Decrease in the use of print journal is obvious because of the fact that the most of the journals are now available in digital platform. Groote & Dorsch (2001) reported the decrease in use of the print collection suggests that many patrons prefer to access journals online.

Table 8 elaborates how often the students log on to the social networking site. 29% male and 36% female log on constantly. It also seen that the students log on based on their convenience as they differ on the frequency. There are still students who have never use the social networking site.

Table 9 shows the rate of use of social networking sites by the students. They use for one purpose or the other. 34% male and 47% female do not use freebies. Quite often they used to find information. The frequency of use varies.

Table 10 shows that the students who have respondent have at least an account with one social networking sites. The most use social networking sites are the Whatsapp 63% male and 67% female and Facebook 60% male and 61%female followed by Google Plus and Twister. Students need to encourage to open multiple accounts so that they can collaborate and network with a larger audience.

In Table 11 the students were asked to give reason for joining the social network. The main reason was to stay in touch with family and friends, share knowledge with others and also to stay up-to-date with the community. 52% male and 55% female use to stay in touch with their family and friends. While 42% male and 44% female

Table 8. Use pattern of social networking

Sl. No.	Frequency	No. of Respondent	
		Male	Female
1.	Constantly log on	26 (29%)	22 (36%)
2.	Several times a day	17 (19%)	16 (26%)
3.	Once in a few days	22 (24%)	7 (11%)
4.	Once in a week	10 (11%)	4 (7%)
5.	Never	8 (9%)	12 (19%)
6.	No response	2 (2%)	2 (3%)

Table 9. Social networking services used very often

Sl. No	Frequency	Do not use		Very rarely		Quite often		Very often		No response	
		M	F	M	F	M	F	M	F	M	F
1.	Find some information	10 11%	5 8%	24 27%	12 19%	35 39%	28 45%	18 20%	14 23%	3 3%	3 5%
2.	Get opinion	22 24%	6 10%	18 20%	14 23%	18 20%	19 31%	12 13%	11 18%	20 22%	12 19%
3.	Entertain yourself	13 14%	6 10%	15 17%	15 24%	22 24%	18 29%	20 22%	17 27%	20 22%	7 11%
4.	Socialize	14 16%	7 11%	14 16%	9 15%	20 22%	17 27%	20 22%	20 32%	22 24%	9 15%
5.	Stay up-to date	7 8%	4 7%	15 17%	15 24%	22 24%	18 29%	25 28%	15 24%	21 23%	10 16%
6.	Share your experience	17 19%	21 34%	25 28%	15 24%	13 14%	13 21%	15 17%	5 8%	20 22%	8 13%
7.	Get freebies	31 34%	29 47%	18 20%	14 23%	8 9%	5 8%	8 9%	4 7%	25 28%	10 16%

Table 10. Which of the following social networking media you use?

Sl. No.	Social Networking	No. of Respondent	
		Male	Female
1.	Facebook	54 (60%)	38 (61%)
2.	Whatsapp	57 (63%	40 (67%)
3.	Google Plus	24 (27%)	18 (29%)
4.	LinkedIn	11 (12%)	3 (5%)
5.	Wikis	4 (4%)	2 (3%)
6.	Flickr	0 (0%)	2 (3%)
7.	Blogs	5 (6%)	3 (5%)
8.	Myspace	3 (3%)	1 (2%)
9.	Twister	17 (19%)	12 (19%)
10.	Any other /Instagram	8 (9%)	12 (19%)
11.	No response	14 (16%)	10 (16%)

Table 11. Reason for using social networking sites

Sl. No	Purpose	No. of Respondent	
		Male	Female
1.	It is relevant, active and interesting community	24 (27%)	22 (36%)
2.	Stay up-to-date with the community	29 (32%)	29 (47%)
3.	Meetings with like professionals	27 (30%)	6 (10%)
4.	Communication of research output	19 (21%)	11 (18%)
5.	Creating awareness on new methods	16 (18%)	3 (5%)
6.	Stay in touch with family and friends	47 (52%)	34 (55%)
7.	Sharing knowledge with others	38 (42%)	27 (44%)
8.	Improving organizational visibility	10 (11%)	4 (7%)
9.	Reaching out to people to gain valuable ideas	13 (14%)	14 (23%)
10.	Asking questions from professional colleagues	26 (29%)	18 (29%)
11.	No response	17 (19%)	10 (16%)

use to share their knowledge with others and 32% male and 47% female use for staying up- to- date with the community.

From Table 12, it is clear that social media help to exposed to latest knowledge skills and technology as it shows the highest percentage used 83% male and 55% female. 29% female were help in the dissemination of information while 22% male gained more visibility in their research work which show that social media is helping them.

Table 12. Do social media help you?

Sl. No	Purpose	No. of Respondent	
		Male	Female
1.	In finding a mentor	17 (19%)	7 (11%)
2.	Published research work faster.	11 (12%)	5 (8%)
3.	Gained more visibility in my area(s) of research	20 (22%)	13 (21%)
4.	Help in dissemination of information	16 (18%)	18 (29%)
5.	Connect researchers with similar research interest.	16 (18%)	10 (16%)
6.	Exposed to latest knowledge, skills and technology.	50 (83%)	34 (55%)
7.	Able to find institution suitable for research	18 (20%)	11 (18%)
8.	Any other	16 (18%)	14 (23%)
9.	No response	17 (19%)	10 (16%)

CONCLUSION

The need of bioinformatics educational and end-user support service is clear from the above analysis. Librarians need to established training programs that offer valuable services, such as:

- Workshop on the available types of molecular databases and tools.
- Advanced workshop on specialized resources.
- Marketing of library services.
- User support as an ongoing process.

As bioinformatics is gradually gaining roots in the developing countries. Whilst the capacity for bioinformatics research and training is limited, efforts have been made in the last decade to employ bioinformatics techniques in research targeted at local challenges in biomedical science and agriculture. These developments show that when given the needed support, resource-limited countries like Swaziland can contribute to the use of bioinformatics.

REFERENCES

Adithya Kumari H,, Mahadevamurthy, M., & Chandrashekara, J. (2013). *Use of social networking sites among students of engineering college libraries in Mysore city, Karnataka: A study*. Retrieved August 20, 2015 from http://www.lsrj.in/ArchiveArticles

Alpi, K. (2005). *Bioinformatics training by librarians and for librarians: Developing the skills needed to support molecular biology and clinical genetics information instruction*. Retrieved September 15, 2015 from http://www.istl.org/03-spring/article1.html>

Beautyman, W., & Shenton, A. K. (2009). When does Academic Information need stimulate a school-inspired information want? *Journal of Librarianship and Information Science*, *41*(2), 67–80. doi:10.1177/0961000609102821

Catanlano, A. (2013). Patterns of Graduate. Students Information Seeking Behavior a meta-synthesis of literature. *The Journal of Documentation*, *69*(2), 243–274. doi:10.1108/00220411311300066

Cox, B., & Jantti, M. (2012). Discovering the Impact of Library Use and Student Performance. *Educause Review*. Retrieved from http://er.educause.edu/articles/2012/7/discovering-the-impact-of-library-use-and-student-performance

de Groote, S. L., & Dorsch, J. L. (2001). Online journals: Impact on print journal usage. *Bulletin of the Medical Library Association*, *89*(4), 372–378. PMID:11837259

Devadason, F. J., & Lingam, P. P. (1996). *A methodology for the identification of information needs of users*. Retrieved October 20, 2015 from http://archive.ifla.org/IV/ifla62/62-devf.htm

Goodall, D., & Pattern, D. (2011). Academic library non/low use and undergraduate student achievement: A preliminary report of research in progress. *Library Management*, *32*(3), 159–170. doi:10.1108/01435121111112871

Kuhlthau Carol, C. (1993). A principle of uncertainty for information seeking. *The Journal of Documentation*, *49*(4), 339–355. doi:10.1108/eb026918

Kumar, D. (2009). *Information needs of faculty members and research scholars of Chaudhary Charan Singh University: A case study*. Retrieved October 20, 2015 fromhttp://digitalcommons.unl.edu/cgi/viewcontent.cgi?

Messersmith, D. J., Benson, D. A., & Geer, R. J. (2006). *A Web-based assessment of bioinformatics end-user support services at US universities*. Retrieved September 10, 2015 from http://www.ncbi.nlm.nih.gov/pmc/articles/PMC1525314/

Renwick, S. (2005). Knowledge and use of electronic information resources by medical sciences faculty at The University of the West Indies. *Journal of the Medical Library Association: JMLA*, *93*(1), 21–31. PMID:15685270

Sahu, H. K. & Singh, S. N. (2013). Information seeking behavior of astronomy / astrophysics scientists. *ASLIB Proceedings: New Information Perspectives*, *65*(2), 109-142.

Stone, G., Ramsden, B., & Pattern, D. (2011). Looking for the link between library usage and student attainment. *Ariadne*, *67*. Retrieved from http://www.ariadne.ac.uk/issue67/stone-et-al/

Yadav & Tawmbing. (2015). *Use of bioinformatics resources & tools by users of bioinformatics centre in India*. Retrieved October 12, 2015, from http://digitalcommons.unl.edu/cgi/viewcontent

KEY TERMS AND DEFINITIONS

Bioinformatics: The subject which deals with the application of computer technology in solving biological problems.

Information Need: The desire for information.

User Studies: The methods used for the study of information need for a group of user.

Chapter 12
Research, Leadership, and Resource-Sharing Initiatives:
The Role of Local Library Consortia in Access to Medical Information

Reysa Alenzuela
Iloilo Doctors' College, Philippines

ABSTRACT

A consortium is an association of independent libraries and/or library systems established by formal agreement, usually for the purpose of resource-sharing. The needs of special libraries cannot be fully addressed by regional organization because of its wide scope, thus, a consortium for specific group is deemed useful. This book chapter aims to describe the development of a local consortium and its role in building a culture of research, creating dynamic leadership and discussing how resource-sharing scheme goes beyond traditional inter-library loan. Using focus group discussion, the consortium members thresh out issues and concerns where collaborative research, dynamic leadership and resource- sharing pave way to enhance access to medical information.

DOI: 10.4018/978-1-5225-1871-6.ch012

INTRODUCTION

The primary purposes of forming library consortia are to share resources, to achieve a single purpose, and to reduce costs (Kopp, 1998). Additional purposes are to improve quality in services, and to take advantage of the benefits of information technology (Allen & Hirshon, 1998; Jalloh, 2000). Moreover, for medical libraries, quality information is essential for the advancement of science and improved health. Libraries are not limited to providing information and delivering services for better health care. The role of information specialists in special libraries providing medical and allied health resources, facilitating researches, providing assistance to find solutions to growing epidemics through literatures and enhancing public awareness of health is orthogonal to the role of health practitioners.

Iloilo Medical and Allied Health Consortium (IMAHC) is a local consortium established for the purpose of promoting collaboration and enhancing linkages. Local consortium in the perspective of IMAHC is a group of libraries with similar resources, practices, needs and clienteles. There are instances that the needs for unique resources and distinct practices cannot be fully addressed by regional organization because of its wide scope; thus, a consortium for specific group is deemed useful. This book chapter aims to describe the development of a local consortium. While medical libraries only view information delivery and resource-sharing as the avenues to maximize access, this chapter elucidates a new paradigm for developing dynamic libraries providing medical and allied health resources. The authors posit that robust research initiatives, dynamic leadership and innovative resource-sharing scheme that goes beyond traditional inter-library loan can also maximize access to information.

BACKGROUND

Iloilo City is one of the major areas in the Philippine archipelago of more than 7,100 islands. The city is a highly urbanized and the province where it is politically situated is the centre of education. However, the libraries have a long way to go to meet the 21st century concept that maximize utilization and access of the vast information superhighway. Libraries are looking for ways to address the colossal challenges and one avenue identified to address the issue is through creating a network where resources can be shared and concerted efforts among information professionals can benefit all the members.

A consortium is an association of independent libraries and/or library systems established by formal agreement, usually for the purpose of resource-sharing (Reitz, 2016). In the Philippines, library consortium dates back to the successive establishment of three consortia: the Academic Libraries Book Acquisition Services Association (ALBASA) in 1973, the Inter-Institutional Consortium (IIC) (now South Manila Inter-Institutional Consortium) in 1974, and the Mendiola Consortium (MC) in 1975 (Fresnido & Yap, 2014). Several other consortia spread in three major regions of the Philippine Archipelago.

In the Visayas Region, specifically in Iloilo City, the Inter-Institutional Consortium of Libraries started in February 21, 1980 with an institutional meeting of proposed Science Consortium that paved way to the birth of Inter-Institutional Consortium Libraries for Iloilo City through a Memorandum of Agreement. The members- institution were composed of University of San Agustin, Central Philippine University, University of the Philippines-Visayas and West Visayas State College. This group of academic libraries agreed to make available their library resources and services or instruction and research and other related academic activities for university libraries participating in institutions. In more recent years, three more institutions participated including Western Institute of Technology, John B. Lacson Foundation Maritime University and St. Paul University Iloilo.

In the 90s, dispersion has become widespread demonstrating the realization of the power of collaboration. Consortia have played an important role integrating technologies and amplifying resource-sharing. DOST-ESEP Library network is the first library network providing connectivity by means of the information highway in the Philippines, the (PHnet) building library resources and services in eight academic libraries utilizing technology (David, 1996). Another big leap was the creation of PAARLNET, with more than 150 members scattered all over the country. This is the biggest and widest national academic and research consortium of libraries consortia in the country. Moreover, on consortium based on geographic location, aside from IICL which has been mentioned above, Intramuros Library Consortium (ILC) materialized in 2001 with six member libraries – the Mapua Institute of Technology, Pamantasan ng Lungsod ng Maynila, Colegio de San Juan de Letran, Lyceum of the Philippines, Department of Labor and Employment and the Manila Bulletin. Intramuros Consortium is just one of the many consortia established in the Northern region of the country. In the Southern part of the country (Mindanao region), Davao Colleges and University Network (DACUN) was officially established in 2004, to assist academic, research and extension agenda mapping out a workable arrangement that would facilitate networking among its member-institutions. DACUN is composed of ten universities, namely, Assumption College of Davao, Brokenshire College, Davao Doctors College, Holy Cross of Davao College, The Philippine Women's College of Davao, Rizal Memorial Colleges, University of the Immaculate Conception, University of Mindanao, University of Southeastern Philippines, and the University of the Philippines in Mindanao.

Generally, these consortia has reduced cost of e-resources acquisitions, shared staff skills and expertise and provided common interface to resources. Consortia bring economy, efficiency and equality in information availability and use. Participant institutions in a consortium have access not only to their own resources but sources in the other institutions as well. From previous studies consortia played a significant role in allowing the gap between information resource rich libraries and those, which are resource deficient to be bridged (Pandian et al., 2002).

The study of Fresnido and Yap (2014) identified the activities undertaken by selected academic library consortia in the Philippines as follows:

- Cooperative collection development,
- Coordinated purchasing,
- Inter-library lending,
- Shared cataloguing, co-operative cataloguing and building of online union catalogues or virtual catalogues, etc.,
- Sharing of storage facilities,
- Sharing of human resources,
- Promotion of professional development,
- Sharing of expertise on library automation, networking, digitization, managing digital information assets, etc.,
- Collective lobbying on national and international issues like pricing, copyright, etc.,
- Joint preservation and archiving activities for print and digital material,
- Support in the development of important local or historical collection,
- Digitization of valuable and rare collection,
- Collective promotion, marketing, and publicizing of library services,
- Prosser (2005), specifically on access to information, sees several possible roles for consortia,
- Creating infrastructure and standards for repositories,
- Negotiating article processing charges,
- Developing digital publishing tools,
- Supporting national and local journals turning to Open Access,
- Retrospective digitization of journals, and
- Supporting development of Open Access dissemination of non-journal material.

Library consortia are challenged to take on new roles in a changing scholarly communication landscape. Despite the clear-cut roles, certain factors have hampered the full utilization such as issues on lack of funding, commitment of members, weak leadership, conflicting priorities, conflict among members, geographic barriers, extent of support from administration, varying level of technological development. As noted in the study of Fresnido and Yap (2014) while majority of the surveyed consortia have assessed themselves to be successful, it is evident that there is lack of congruence between the consortia's objectives and undertakings. To enable them to seriously fulfill their missions, they should remain focused and should exert conscious effort in accomplishing their goals.

Focus is the state or quality of having or producing clear visual definition. It is an emphasis on a particular need. The homogeneity of membership partakes in a more effective and responsive way of addressing issues. Like any other special libraries, medical libraries are catering to a particular group of users. Thus, with the above issues facing medical libraries, a local consortium was established in Iloilo City.

THE CONCEPTION OF A LOCAL CONSORTIUM

In the Philippines, the Medical and Health Library Association of the Philippines (MAHLAP) was conceived in 1988. The presence of a national organization specifically for health and medical information providers, the librarians or information specialists has helped in the promotion of professional development. However, a national organization catering about two hundred members cannot address the issue on boosting leadership, creating thought-out strategies that will maximize access to medical information in the region, implementing research initiatives and other concerns on the regional or local level.

Thus, in December 2014, a group of academic libraries with special collection and/ or separate libraries for medical and allied health courses agreed to share resources to amplify their resource capability and maximize its utilization. The issue of continuing education program which are not fully availed due to geographic barriers was also raised. Problems on succession and transfer of expert and technical knowledge were other concerns the group has identified. Librarians from medical libraries also deemed it necessary to establish linkages to strengthen research, print and non-print resources and ICT capability. Thus, the group decided to build on their capacity and update their libraries through opening its doors to other libraries and information centers having similar needs. Along with these thrusts, the first local Consortium for medical and allied health libraries and information center was established dubbed as Iloilo Medical and Allied Health Consortium or IMAHC.

IMAHC is highlighting the role of local consortium beyond traditional library practices working on a networked paradigm where member–institutions capitalize on their distinct capabilities and the homogeneity of their goals. The local consortium is working on three pillars- research, leadership and resource-sharing. Figure 1 provides an illustration.

Figure 1. The three-pronged approach for local consortium

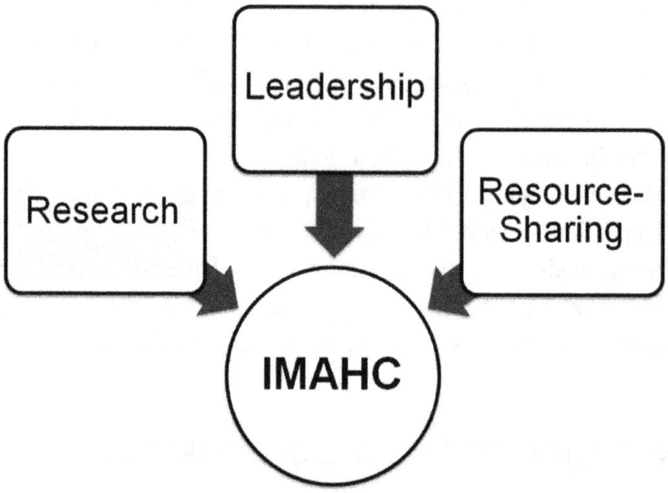

As to leadership, recognizing the diverse membership from seasoned library managers to early career professionals, the group is working on sharing technical knowledge to create dynamic and well- organized programs that will build on the capability of members and non-member medical libraries. As standardized practices in terms of technical services which can only be found in a special library cannot be addressed by national organizations, adoption of unique standards through the technical knowledge of consortium members is also proposed. Moreover, IMAHC endeavors in developing a culture of research by crafting a research agenda emphasizing local readership, local research publications and grey literatures produced in the region. The research initiative is working on the impact study of local and international publications produced by local authors in the medical field that will enable the users to utilize grey literatures and local studies. Likewise, it aimed to create standardized practices in the region that are deemed beneficial to the end users. On the third pillar, recognizing that effective utilization of ICT infrastructure and resources requires the expertise to execute and implement projects relating to innovative technology, the role of resource-sharing is explained. This approach of local libraries gears towards one common goal- maximizing access of medical information.

Leadership: Exchange of Information and Enhanced Cooperation

Leadership roles for managing people, system and resources and educating health information professionals in a changing environment are critical. The former President of MAHLAP, Elnora Conti, noted:

Medical and health professionals through the libraries and information centers in the academe or hospitals access to these information technologies and use these technologies in their practice resulted to the fast and efficient delivery of medical and health services. There is no doubt that the symbiotic relationship of medical and health professionals and the information handlers - the librarians, or information resource specialist will lead to a high level of health delivery service. (Conti, 2011)

IMAHC was formed with six member libraries:

1. Iloilo Doctors' College (IDC),
2. Integrated Midwives Association of the Philippines (IMAP) Foundation School of Midwifery,
3. PHINMA- University of Iloilo (PHINMA-UI),
4. St. Paul University- Iloilo (SPU-Iloilo),
5. University of San Agustin (USA), and
6. Western Visayas State University (WVSU) College of Medicine Library.

The health and medical information providers, the librarians or information specialists are working hand in hand with the medical professionals. But it takes dynamic leadership and exemplary ability to address the ever growing demands in providing quality information. In a focused group discussion, the members of IMAHC sorted out issues faced by librarians; some are their distinct leadership roles working in academic libraries that provide medical and allied health resources.

From the above issues and concerns, IMAHC intended to lead towards the following initiatives:

1. Resource-sharing,
2. Capacity building,
3. Standardized practice,
4. Preservation of Library Collection,
5. Technology,
6. Research.

Table 1. Issues in medical libraries and its implications on leadership roles

Issues	Implications on Leadership Roles
Management of leadership turn-over	Library managers must have vision for the library and the organization/ institution served. Every library manager must instill individual and organizational readiness for change.
Enhancing competencies through cost-effective localized capacity building programs	A library leader must have passion to share knowledge, skills and experience.
Cooperative collection development	Every library manager must have the power to convey the importance of cooperative collection development.
Coordinated purchasing/ acquisition to maximize resources	Library managers must create thought out strategies for coordinated purchasing/ acquisition.
Building of online resources	Librarians must instigate ways of expanding knowledge base of medical resources.
Address current and prevailing issues (e.g. slow internet access and uneven access to medical information)	Librarians must have the persuading charisma to convince administrators to capitalize on technology. Librarians must use their concerted effort to facilitate access to medical information if the said information is not available to one member-library due to ICT- related problems.
Sharing of technical expertise to be able to adopt to standardized practices	Library leaders must motivate colleagues with technical knowledge to transfer/ share knowledge through trainings.
Development of important local or historical collection	Library managers must build on winning the confidence of individuals and organizations who are primary producers of local/ historical information or resources to entrust to them the preservation of valuable local/ historical resources.
Institutionalizing standardized practices for more efficient access and retrieval	Library managers and information professionals must have sufficient knowledge and capability to deal with complex policy issues that may arise in Institutionalizing standardized practices.
Developing competencies to adopt to ASEAnization	Keeping abreast with global issues and regional concerns. Equip staff and colleagues to be able to handle challenges and changes.
Facilitate fast and efficient delivery of medical and health services through information support	Oversee and identify the technical knowledge of staff that can be instrumental in facilitating fast and efficient delivery of medical and health services through information support.

From traditional tasks of managing information and services, librarians are now called to more engaging and dynamic role of being directly involved creation, development and dissemination of information. Also, the changing needs and behavior of users are the driving force of libraries to re-think and re-design its roles, functions and services beyond the traditional means. It has been a cry to begin embracing technology and initiate a culture of research.

RESEARCH: REDEFINING ACCESS TO MEDICAL INFORMATION

Iloilo Medical and Allied Health Consortium (IMAHC) with its member institutions IDC, IMAP, PHINMA-UI, SPU Iloilo, USA and WVSU agrees to advance the promotion of a culture of research by encouraging library research cooperation and sharing of institutional researches. IMAHC members share information resources, facilities, knowledge and skills to facilitate such coordination and integration of their activities for the purpose of carrying out this thrust. Thus, IMAHC has raised these research agenda.

At the heart of the research agenda are the principles that guide research decisions. These principles complement and align with the goals of the Consortium:

- Support innovative and flexible structures to develop a team-based research;
- Enhance awareness, utilization and preservation of local knowledge, regional best practices and dynamic collaboration; and
- Document and publish the process of creating collaborative research in medical libraries in the City (Iloilo).

The scarcity of time and lack of institutional support are some of the great challenges faced by libraries in this city. Confidence and exposure of academic librarians in developing research is another issue that hampers the development of collaborative researches. Conquering this challenge depends on the ability to create and support core capabilities and commit resources to initiate this tedious and daunting initiative. Within the core values, principles and awareness of the challenges, IMAHC works within its strengths and challenges to advance the following prospective research topics:

1. A developmental study for adopting consortia Demand-Driven Acquisition (DDA),
2. Institutionalizing more precise searches through use of Medical Subject Headings (MeSH),
3. Shared preservation/retention initiatives,
4. Study on the use of internet among Medical Practitioners and Health Care Professionals,
5. Production, Dissemination and Use of Medical Grey Literatures in Libraries.

It can be gleaned that one aspect of research is creation of standardized practices. From DDA to MeSH, the goal of the consortium is to advance knowledge to institutionalize practices that will maximize access of users and enhance their search experience. Moreover, user study, preservation and access are identified as the priority for the coming years. The group hopes that availability of empirical studies will redefine and re-engineer the way information is distributed.

Resource-Sharing: Forging Cooperation

Integration is the by-word nowadays. The beauty of collaboration, cooperation, resource-sharing and networking is valued. Librarians play an important role in any aspect of knowledge acquisition as they have the expertise in information retrieval and delivery. Librarians can assist faculty researchers without the limitation of time and space. Through chat rooms or traditional emails, delivery of literatures for researches can be facilitated.

This paper advances that effective resource- sharing is an offshoot of interactions of various dynamic factors. In the focused group discussion and needs assessment conducted by this group, leadership roles and research initiatives emerged as the major stirrers that facilitate successful resource-sharing and collaboration. A summary of the results of discussion is provided in Table 2.

Quality information is must in knowledge and evidenced based health care system. Health sciences librarians needs to take part in more dynamic roles. Consortia can bring economy, efficiency and equality in information availability and use. Participant institutions in a consortium have access not only to their own resources but sources in the other institutions as well. This can allow the gap between information resource rich libraries and those which are resource deficient to be bridged (Pandian, et al., 2002). Most of the resource-sharing schemes follow reciprocity of benefits. However, IMAHC members deemed that a new platform should be ventured. Some members which were included do not have sufficient resources; hence, from the inception, it is very clear these small academic institutions will play more as recipients. The quantity of resources was deemed irrelevant as most of the resources shared are non-print materials. The only major challenge encountered was

Table 2. Results of focus group discussion on factors in the success of resource- sharing

Possible Solutions	IDC	IMAP	PHINMA-UI	SPU-I	USA	WVCST	n
Strong leadership of librarian	✓	✓	✓	✓	✓	✓	6
Conduct research	✓	x	x	✓	✓	✓	4
Plan	x	✓	✓	x	x	✓	3
Communicate with administrators	✓	x	x	x	x	x	1
Involvement of all stakeholders	✓	x	x	✓	x	x	2

convincing non-librarian administrators on a new platform of resource-sharing. In this case, clear communication and trust on the leadership of the librarian is significant.

It is clear that the creation of the web and networks brought many possibilities for integrating consortia activities both in developed and developing countries. The library consortia are shifting from a peripheral and limited position of resource-sharing to an integrated system-wide resource-sharing in recent years in the West (Talawar, 2009). Resource-sharing is considered to be a great advantage of consortia for libraries, as today, the ability for users to access resources is often more important than collection building within a particular library. Through a library consortium, the collective strength of resources of various institutions available to it can be increased. The consortia enable libraries to gain the benefits of wider access to electronic resources at an affordable cost (Singh and Singh, 2004). On research, evaluation of the effectiveness of this program is not the only data needed. A study on user behavior, information retrieval and effectiveness of policies must also be implemented. The knowledge society continuously metamorphoses. Hence, research is an important pillar that should be carried out to measure objectively the success of programs.

THE TASKS AHEAD

This local consortium is the first network of academic libraries providing medical and allied health information developed in the country. Apparently, the projects are on-going and further assessments are needed. Amplifying access to medical information means exhausting all information available. The internet is just a tip of the iceberg. Librarians must endeavor to look into local studies and grey literatures that academic institution and research centers develop. Exchange information and enhanced cooperation in libraries is a must to provide the best information possible. This is also very relevant as to grey literatures. Access to medical information implies a colossal task: leading the digitization of literatures in other regions, teaching librarians to cull and preserve even providing assistance in the translation. Research, leadership and resource- sharing initiatives play an important role for the Consortia, and ultimately, the goal is to maximize access to medical information.

ACKNOWLEDGMENT

Acknowledgement for the data on history of Inter-Institutional Consortium Libraries for Iloilo City to Ms. Regina Maligad, Director of Libraries, University of San Agustin.

REFERENCES

Allen, B. M., & Hirshon, A. (1998). Hanging together to avoid hanging separately: Opportunities for academic libraries and consortia. *Information Technology and Libraries*, *17*(1), 36–44.

Conti, E. (2008). *Ten years of service to medical and health librarianship*. Retrieved January 22, 2015 from http://www.mahlap.org/index.php?option=com_content&task=view&id=3&Itemid=20

Fresnido, A. M. B., & Yap, J. M. (2014). Academic library consortia in the Philippines: Hanging in the balance. *Library Management*, *35*(1/2), 15–36. doi:10.1108/LM-04-2013-0028

Homan, J. M., & McGowan, J. J. (2002). The Medical Library Association: Promoting new roles for health information professionals. *Journal of the Medical Library Association: JMLA*, *90*(1), 80–85. PMID:11838464

Jalloh, B. (2000). A plan for the establishment of a library network or consortium for Swaziland: Preliminary investigations and formulations. *Library Consortium Management: An International Journal*, *2*(8), 165–176.

Kopp, J. (1998). Library consortia and information technology: The past, the present, the promise. *Information Technology and Libraries*, *17*(1), 7–12.

Nfila, R. B., & Darko-Ampem, K. (2002). Developments in academic library consortia from the 1960s to 2000: A review of literature. *Library Management*, *23*(4/5), 203–212. doi:10.1108/01435120210429934

Pandian, P. M., Jambhekar, A., & Karisiddappa, C. R. (2002). IIM digital library system: Consortia-based approach. *The Electronic Library*, *20*(3), 211–214. doi:10.1108/02640470210432357

Prosser, D. (2005). The economics of Open Access. In *ICOLC autumn 2005 meeting*. Retrieved from http://www.pfsl.poznan.pl/icolc/1/ICOLC_ Econ.ppt

Reitz Joan, M. (Ed.). (2016). *Online Dictionary for Library and Information Science*. Retrieved January 20, 2016 from http://www.abc-clio.com/ODLIS/odlis_c.aspx

Singh, S., & Singh, S. (2004). Need for joining library consortia: a study of Vikram University Library. In *Proceedings of National Seminar on Library Consortia*.

Smith, I. (2005). Achieving readiness for organisational change. *Library Management*, *26*(6/7), 408–412. doi:10.1108/01435120510623764

Talawar, G. G. M. V. G. (2009). Library consortia in developing countries: An overview. *Program, 43*(1), 94 – 104. DOI:10.1108/00330330910934138

Taole, N., & Dick, A. L. (2009). Implementing a common library system for the Lesotho Library Consortium. *The Electronic Library*, *27*(1), 5–19. doi:10.1108/02640470910934551

KEY TERMS AND DEFINITIONS

Consortium: An association of independent libraries and/or library systems established by formal agreement for the purpose of resource sharing.

Demand-Driven Acquisition: A kind of acquisition where title (print or non-print) is acquired as triggered for purchase via significant use by patrons as defined by the library.

Iloilo City: A highly urbanized city located in the Western Visayas region of the Philippine archipelago.

MeSH: A comprehensive controlled vocabulary for the purpose of indexing journal articles and books in the life sciences, it serves as a thesaurus that facilitates searching.

Philippines: A Southeast Asian country in the Western Pacific, comprising more than 7,000 islands. It is divided into three major regions, namely: Luzon (Northern), Visayas (Central) and Mindanao (Southern).

Resource-Sharing: An initiative done by libraries to share resources with other libraries through a formal agreement and usually as part of a consortium or partnership.

Chapter 13
Information Retrieval and Access in Cloud

Punit Gupta
Jaypee University of Information Technology, India

Ravi Shankar Jha
Jaypee University of Information Technology, India

ABSTRACT

With increase of information sharing over the internet or intranet, we require techniques to increase the availability of shared resource over large number of users trying to access the resources at the same time. Many techniques are being proposed to make access easy and more secure in distributed environment. Information retrieval plays an important to serve the most reliant data in least waiting, this chapter discuses all such techniques for information retrieval and sharing over the cloud infrastructure. Cloud Computing services provide better performance in terms of resource sharing and resource access with high reliability and scalability under high load.

INTRODUCTION TO CLOUD COMPUTING

Describe Cloud computing is a trending topic that many find confusing but It isn't, though, in fact, most of those who claim not to understand the subject are part of the majority that uses it daily. In basic terms, cloud computing is the word used to describe different scenarios of computation in which computing resource is delivered as a service over a network connection (usually, over the internet). Datacenter

DOI: 10.4018/978-1-5225-1871-6.ch013

hardware and software that the vendors use to offer the computing resources and services. Cloud technology allows for the automatic provision and releases resources as per requirement and when it is necessary, thus ensuring of resource availability match to current demand as possible. This is a defining characteristic that completely differentiates it from other computing models where the resource is delivered in blocks (e.g., individual servers, downloaded software applications), usually with fixed capacities and high costs. See more characteristics in Figure 1. With cloud computing, the end user usually pays only for the resource they use and so it avoids the inefficiencies and expense of any unused computation models (Sakr, 2010).

However, the advantages of cloud computing are not limited to flexibility even there are many things which make it more reliable and make a very good choice of computation option. Enterprise industries can also benefit (in varying degrees) from the economies of scale created by setting up services all together with the same computing environments, and the reliability of physically hosting services across multiple servers which may be like geographically on same machine or differentplace where individual system failures do not break or affect the continuity of the service (Sen, n.d).

Over conventional method of resource sharing over a private server fails to provide Quality of Service as assured under large users or requests made for a resource. Cloud computing is a solution to guarantee an assured Quality of Service in any conditions of high load or sharing large resources.

Figure 1. Characteristics of cloud computing

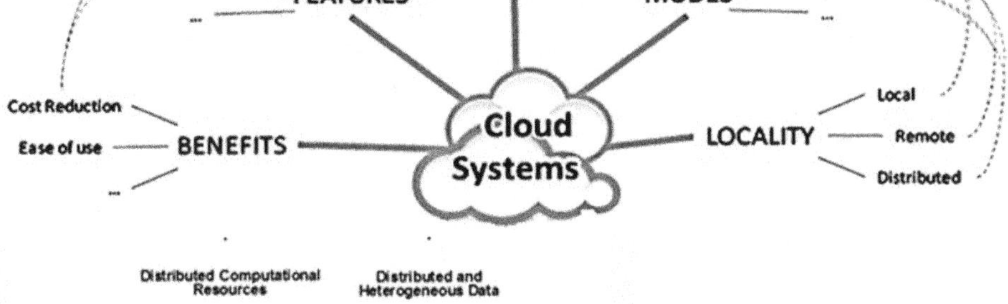

It has been categories in major parts naming as public, private and hybrid cloud computing. To understand these three terms, let's imagine it in your mind to have a clear vision. See Figure 2.

Public Cloud

A public cloud, for example, is a cloud in which services and infrastructure are hosted off-site or we can say over the server by a cloud provider or can say a vendor, shared across their client base and accessed by these clients via public networks or the internet. Public clouds offer great economies of scale and redundancy but are more vulnerable than private cloud setups due to their high levels of accessibility and services management.

Private Cloud

Private clouds on the other hand use pooled base services and infrastructure stored and maintained on a private network like fully dedicated to specific client– whether physical or virtual – accessible for only one client. The obvious benefits to this are greater levels of security and control. It is not good to keep the valuable resource in public. it is costly but good for as the enterprise or client in question will have to purchase or rent and maintain all the necessary software and hardware.

Figure 2. Type of cloud

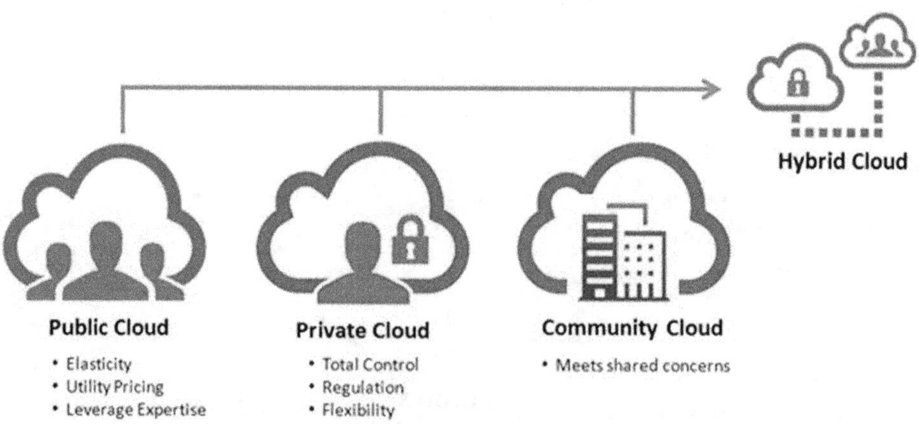

Hybrid Cloud

The final cloud option is a hybrid cloud and this, as the name suggests, combines both public and private cloud elements. A hybrid cloud allows a vendor or cloud provider to maximize their efficiencies, by utilizing the public cloud for non-sensitive operations while using a private cloud setup for sensitive or high valuable and to perform critical operations, companies can ensure that they are paying for what is necessary. if we move broadly to the development and programmatic manner then there are 3 models or pillar of cloud computing which describe the service on offer, these are Software as a Service (SaaS), Platform as a Service (PaaS) and Infrastructure as a Service (IaaS) (Firdhous, Ghazali & Hassan, 2011). See this layered cloud architecture in Figure 3

Infrastructure as a Service (IaaS)

This is one of the basic three fundamental service models of cloud computing alongside Platform as a Service (PaaS) and Software as a Service (SaaS). As with all cloud computing services it provides access to computing resource in a virtual environment, across an internet or public connection. In the case of IaaS the computing resource provided user-specific on virtualized hardware, in other words, computing infrastructure. The definition includes such offerings computing components as virtual server space, network connections, bandwidth, IP addresses and load balancers. In other words, the computation components are reserved for user basis. These components physically lied on same or different places geographically but,

Figure 3. Layered architecture

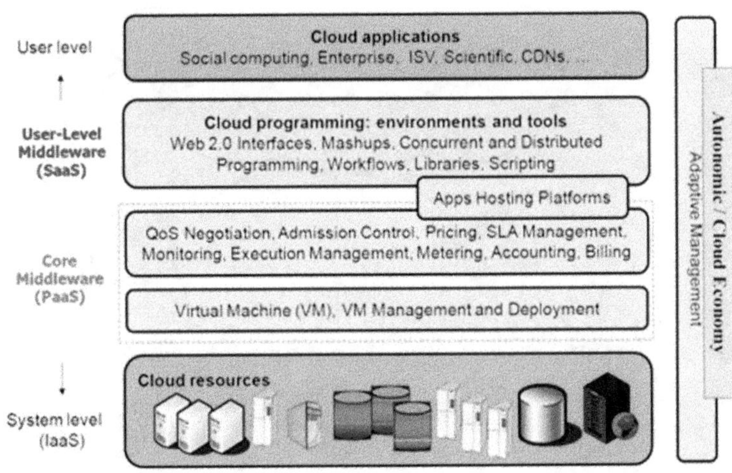

the pool of hardware resource is pulled from a multitude of servers and networks usually distributed across multiple data centers, all of which the cloud provider or vendor is responsible for maintaining. The client, on the other hand, is given access to the virtualized components in order to build their own IT platforms and use them as per there needed. with comparison to other two forms of cloud hosting, IaaS can be utilized by enterprise customers to create cost-effective and easily scalable IT solutions which imply that as per customers need to expand the resource or install some components then it can be done by upgrading subscription with generating a request to vendor and it will their cost to use it.

It is having high benefits, for example, system failures may possible in a single server or in central approaches but in the cloud if one server or network switch, were to fail, the running service would be unaffected due to the remaining multitude of hardware resources and redundancy configurations. For many services if one entire data center were to go offline, never mind one server, the IaaS service could still run successfully, as illustrated in Figure 4.

Platform as a Service (PaaS)

This is basically for enhancing computation and productivity of an enterprise. Platform as a Service, often simply referred to as PaaS, is a category of cloud computing that provides a platform and environment to allow developers to build

Figure 4. Cloud infrastructure as a service (IaaS)

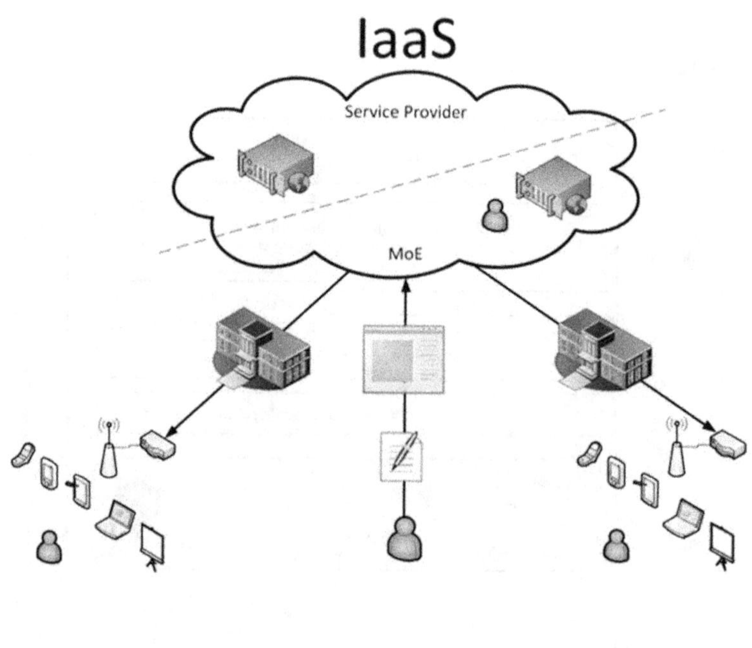

their own applications and services over the internet. See Figure 5. PaaS services are hosted in the cloud and accessed by users simply via their web browser or can say remote accessibility.

Platform as a Service allows users to create software applications using tools supplied by the vendor. PaaS services can consist of pre-configured features as client demand or customers can subscribe to, they can choose to include the features that meet their requirements while discarding the rest off as unused. Consequently, packages can vary from offering simple point-and-click frameworks where no client expertise is required to supplying the infrastructure options for advanced development. The infrastructure and applications as per subscription are managed for customers and support is available. Services are constantly updated, with existing features upgraded and additional features added. PaaS providers can assist developers from the conception of their original ideas to the creation of applications, and through to testing and deployment. This is all achieved in a managed mechanism in PaaS. Software developers, web developers, and businesses can benefit from PaaS. Whether building an application or service which they are planning to offer over the internet or software to be sold out of the box, software developers may take advantage of a PaaS solution. For example, web developers can use individual PaaS environments at every stage of the process cycle to the developer, test and finally host their websites. However, businesses that are developing their own internal software can also utilize Platform as a Service, particularly to create distinct

Figure 5. Cloud platform as a service (PaaS)

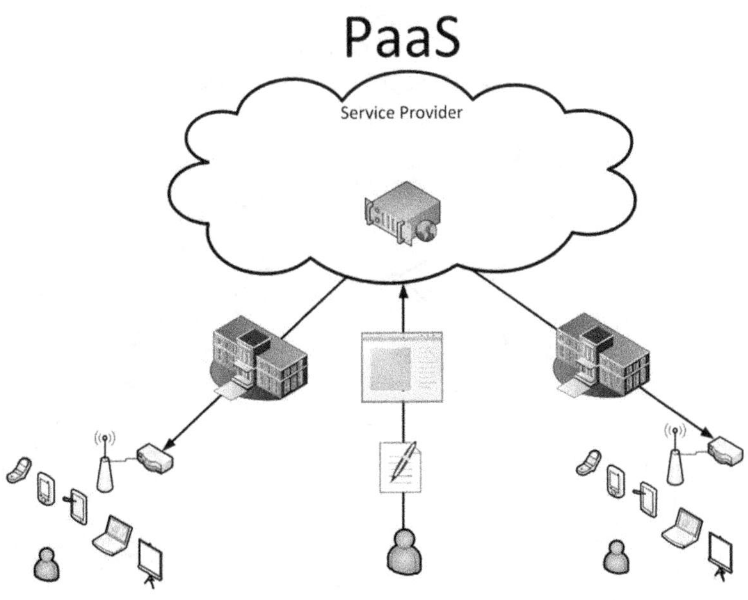

bounded development and testing environments which allow thereal owner to manage their services specifically.

Software as a Service (SaaS)

SaaS, or Software as a Service, describes any cloud service where consumers are able to use their needy software applications over the internet. The applications are hosted in the cloud and can be used for a wide range of tasks for both individuals and organizations. See Figure 6. Google, Twitter, Facebook, and Flickr are all examples of SaaS, with users able to access the services via any internet enabled devices like smartphone or PDA. Enterprise users are able to use applications for a range of needs, including accounting, scientific and invoicing, tracking sales, planning, performance monitoring and communications (including webmail and instant messaging). SaaS is often referred to as software-on-demand and utilizing it is similar quality to renting software rather than buying it. With traditional software applications, you would purchase the software as a package and then install it onto your computer. The software's license may also limit the number of users or devices where the software can be deployed. Software as a Service user, however, subscribe to the software instead purchase it, usually on a monthly basis. Applications are purchased and used online with files saved in the cloud storage rather

Figure 6. Cloud software as a service (SaaS)

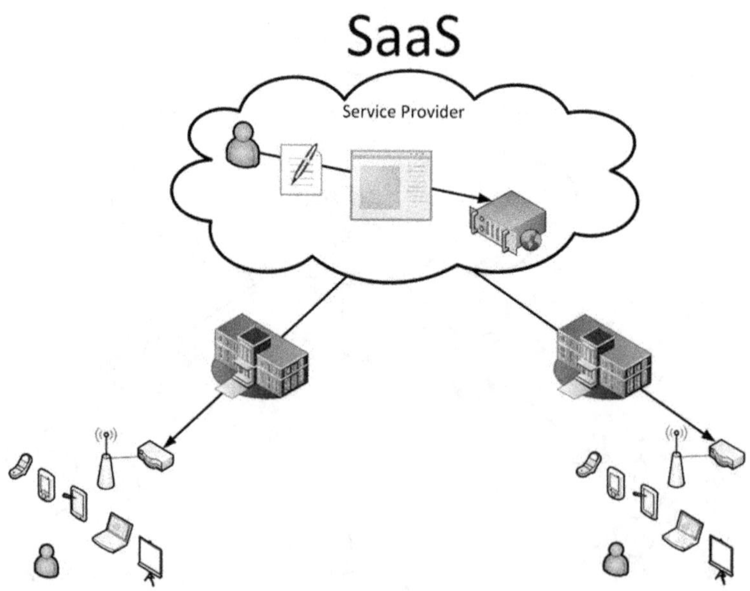

than on individual computers make data file all-time availability anywhereanytime (Afgan et al. 2012)

Office software is the best example of businesses utilizing SaaS. Tasks related to accounting, scientific calculation, invoicing, sales and planning can all be performed through Software as a Service. Businesses may wish to use one piece of software that performs all of these tasks or several that each performs different tasks which make it more complicated. The required software can be subscribed to via the internet and then accessed online via any computer in the office using a username and password. If needs change, they can easily switch to software that better meets their requirements. Everyone who needs access to a particular piece of software can be set up as a user, whether it is one or two people or every employee in a corporation that employs hundreds, SaaS deals with this.

SaaS are just a scalable version of web services been hosted over the servers to fulfill client requests. Web services best example can be IEEE, ACM and Springer digital library which uses web services over the cloud to provide resources without failure.

WEB SERVICES

A Web service is an abstract notion that must be implemented by a concrete agent which deals both side client and server side or we can say an interpreter for them. The agent is the concrete piece of software or hardware that sends and receives messages in the middle, while the service is the resource characterized by the abstract set of functionality that is provided. To make it better understand, we service may implement in a specific language using an agent and after some time we use some different programming language with the same functionality it implies that agent got changed but services are same. The main idea behind web service is to provide some functionality on behalf of it's owner or a person or organization, such as an individual or business. The provider is person or organization that provides an appropriate agent implement a specific service and a requesting entity is a person or an organization that wants to use that service. It will user a requester agent to access functionality of web service, requester agent interacts with service provider agent and exchange messages to communications (Figure 7).

The mechanism of the message exchange is documented in a Web service description (WSD). As we can see in the web-service architecture as in Figure 7. The WSD is a machine-processable specification manifesto of the Web service's interface, written in WSDL. It defines the message formats, datatypes, transport and communications protocols, and transport serialization formats that should be used between the requester agent and the service provider agent. It also specifies one or

Figure 7. Web service architecture

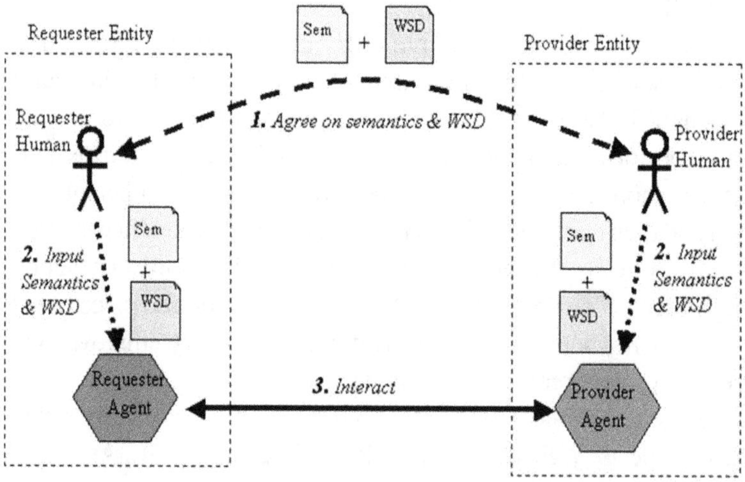

more network locations at which a provider agent can be invoked or may provide some information about the message exchange pattern that is expected. In essence, the service description represents an agreement governing the mechanics of interacting with that service.

The meaning of a Web service is the shared expectation about the behavior of the service, in particular in response to messages that are sent to it. In effect, this is the contract between the requester entity and the provider entity regarding the purpose of the interaction. Although this contract represents the overall agreement between the requester entity and the provider entity on how and for what their respective agents will interact, it is not necessarily written or explicitly negotiated. It may be explicit, machine processable or human-oriented, and it may be a legal agreement or an informal agreement.

Working

There are many ways of service or access web-service but in general, these steps are followed by web-service.

1. **Step One:** To initiate interaction of both side agents, first they have to know each other or atleast one of them know another. Here to access the web-service provided by the provider can be accessed parallel via multiple agents which are known to them via skeleton on the client side to initiate interaction.

2. **Step Two:** The requester and provider entities somehow agree on the service agreement and semantics that will govern the interaction between the requester and provider agents.
3. **Step Three:** The service description and semantics are serialized by the requester and provider agents to begin service.
4. **Step Four:** The requester and provider agents communicate via messages, thus performing some task on behalf of the requester and provider entities. In details, the exchange of messages with the provider agent represents the concrete manifestation of interacting with the provider entity's Web service.

The technologies that we consider here, in relation to the Architecture, are XML, SOAP, WSDL. However, there are many other technologies that may be useful. XML solves a key by offering a standard, flexible and inherently extensible data format, XML significantly reduces the burden of data deploying the many technologies needed to ensure the success of Web services. SOAP (Simple Object Access Protocol), provides a standard, extensible, composable framework for packaging and exchanging XML messages. WSDL(Web Service Description Language) is a language for describing Web services. It describes Web services starting with the messages that are exchanged between the requester and provider agents. As shown in Figure 7 and mentioned in the steps, engaging the web-service is as below in extended form:

1. In the first step, the requester and provider entities become known to each other, after that in the sense that whichever party initiates the interaction must become aware of the other party. There are two possible cases:
    a. In this case, the requester agent will be the initiator. In this case, we would say that the requester entity must become aware of the provider entity, the requester agent must somehow obtain the address of the provider agent to exchange the messages.
    b. There are two ways this may typically occur:
        i. The requester entity may obtain the provider agent's address directly from the provider entity.
        ii. The requester entity may use a discovery service (Web Service Glossary) to locate a suitable service description via an associated functional description, either through manual discovery or autonomous selection.
    c. In other cases, the provider agent may initiate communications via exchanging the messages between the requester and provider agents. In this case, saying that the requester and provider entities become known to each other actually means that the provider entity becomes aware of the

requester entity, the provider agent somehow obtains the address of the requester agent. How this goes on is application dependent and irrelevant to this architecture. Although this case is expected to be less common than when the requester agent is the initiator, to make it work requester must have subscribed for push scenario.

2. Step two begins after completion of step one, the requester entity and provider entity agree on the service description (a WSDL document) and semantics that will allow the communications between the requester agent and the provider agent. This does not necessarily mean that the requester and provider entities must communicate with each other. It simply means that both parties must have the same (or compatible) understandings of the service description as previously described the format of information which can be understood by them and semantics, and intend to uphold them. There are many ways this can be achieved, such as:
    a. The requester and provider, both entities may communicate directly with each other, to explicitly agree on the service description and semantics.
    b. The provider entity may publish and offer both the service description and semantics as take-it-or-leave-it "contracts" that the requester entity must accept unmodified as per condition of using it.
    c. The service description and semantics may be defined as a standard by an industry organization, and used by many requester and provider entities in parallel. In this case, the act of the requester and provider entities reaching an agreement is accomplished by both parties independently conforming to the same standard.
    d. The service description and semantics may be defined and published by the requester entity, and offered to provider entities on a take-it-or-leave-it basis. This may occur, for example, if a large organization requires its suppliers to provide Web services that conform to a particular service description and semantics. In this case, the agreement is achieved by the provider entity adopting the service description and semantics that the requester entity has published.
3. In step three, the service description and semantics are input to, or possessing in bodily in, both the requester agent and the provider agent as appropriate manner. In other words, the information in them must either be input to or implemented in, the requester and provider agents. There are many ways this can be achieved, and this architecture does not specify or care what means are used. it can be understood with an example. An agent could be hard coded to implement a fixed service description and semantics which serve a specific functionality, or an agent could be coded in a more general way, and the desired service description and or semantics could be input at run time or an agent

could be created first, and the service description and or semantics could be generated or deduced from the agent code. For example, a tool could examine a set of existing class files to generate a service description. Regardless of the approach used in service, from an information perspective, both the semantics and the service description must somehow to be input to, or implemented in, both the agents before they can interact.

4. In the final step, the requester agent and provider agent exchange SOAP messages on behalf of their owners, which interact this and access the service functionality.

In summary, it is convenient and evocative to say that the requester and provider entities must agree on the semantics and the service description that will govern the communication between the requester and provider agents, but it would be more accurate to say that they simply need to have a non-conflicting view of the semantics and service description of the interaction.

CLOUD AND WEB SERVICES

Cloud Software as a service layer is responsible for implementation and maintain the Quality of service of a web service. Whenever a web service is created it is allocated a new virtual machine at IaaS layer in the cloud which is independent and can rescale the resource at any point of time whenever the number of requests increases creating more new instances of web service is a virtual machine (VM). See Figure 8. In the cloud environment, whenever a web service has requested an instance of serve at SaaS level is created and a secure connection is established to the VM to full fill the request.

To maintain the fault tolerance of the service, many replicas of the VM are created after an equal interval of time so as to overcome and start new VM in case of failure without knowing it to the user. This functionality is known as the migration of services and which also helps in making the system scalable and fault tolerant in nature. In library science, main functionality is to serve access to resources 24x7 to the users all over the globe. This functionality can be fulfilled by cloud computing framework with the features and characteristics as discoursed above with high QoS.

Figure 8. Web service architecture in the cloud

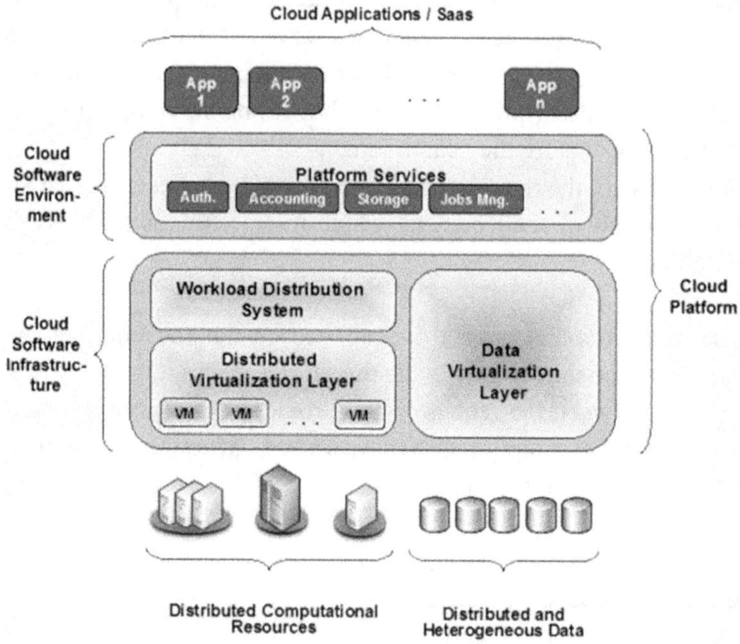

CLOUD COMPUTING AND BIOINFORMATICS

Bioinformatics is a data intensive field of study where the information and computer technology is being used to solve biological problem. The biological data analysis, storage of experimental data, preprocessing and analysis of experimental data is the major tasks undertaken by the bioinformatics community. Managing omics data requires large space for data storing as well as services for data preprocessing, analysis, and sharing. The advancement of computing technology has resulting scenario where bioinformatics tools, often implemented as web services, for the management and analysis of data stored in geographically distributed biological databases (Calabrese & Cannataro, 2016). There are various applications helping bioinformatics services available on cloud platform (Shanahan, Owen & Harrison, 2014). The cloud infrastructure commonly for bioinformatics are Amazon EC2 (available as Infrastructure as a Service (IaaS), Azure, Google AppEngine and Heroku (Leite, Magalhaes & Melo 2012).

In order to finish the large data intensive experiments in less time, parallel computing technology with software application is being used. The bioinformatics applications are also being developed using parallel computing, so called "Cloud

Computing" technology and many applications are now being run on Cloud. Schatz, Langmead & Salzberg, (2010) reviewed various bioinformatics applications which are being run on Cloud (Table 1).

CONCLUSION

This chapter discusses cloud computing and its role in effective and efficient data retrieval and management. We have discussed various services provided by cloud providers at infrastructure level, platform level and service level for public, private or hybrid cloud environment. Cloud computing helps us to overcome the disadvantages of conventional web server based services and user need to pay for only those services which are been used and need not require a large infrastructure to deploy the services. Cloud computing based web services which are well known in

Table 1. Bioinformatics applications available on the cloud

Applications	Description
CloudBLAST (Matsunaga et al., 2010)	Scalable BLAST in the cloud (http://www.acis.ufl.edu/%7Eammatsun/mediawiki-1.4.5/index.php/CloudBLAST_Project)
CloudBurst (Schatz et al. 2010)	Highly sensitive short-read mapping (http://cloudburst-bio.sf.net)
Cloud RSD (Wall et al. 2010)	Reciprocal smallest distance ortholog detection (http://roundup.hms.harvard.edu)
Contrail	*De novo* assembly of large genomes (http://contrail-bio.sf.net)
Crossbow (Langmead et al. 2009)	Alignment and SNP genotyping (http://bowtie-bio.sf.net/crossbow/)
Myrna	Differential expression analysis of mRNA-seq (http://bowtie-bio.sf.net/myrna/)
Quake	Quality guided correction of short reads (http://github.com/davek44/error_correction/)
AWS Public Data	Cloud copies of Ensembl, GenBank, 1000 Genomes and other data (http://aws.amazon.com/publicdatasets/)
CLoVR	Genome and metagenome annotation and analysis (http://clovr.igs.umaryland.edu)
Cloud BioLinux	Genome assembly and alignment (http://www.cloudbiolinux.com/)
Galaxy (Giardine et al., 2005)	Platform for interactive large-scale genome analysis (http://galaxy.psu.edu)

Source: Schatz, Langmead & Salzberg, 2010

industry for data retrieval and storage helps user to establish and launch a service on the fly, with high reliability and scalability. Moreover, cloud takes care of everything i.e. the services provided by cloud assures least cost and high reliability under high request load or under loaded condition by scaling the resources using replica management techniques for high availability. The importance of cloud computing is increasing day by day. Various subjects and areas extending their services through cloud computing. Bioinformatics too using various application on cloud and new horizon are being established due to the advancement of the cloud computing. At the end, we would like to conclude that cloud computing has changed the way, data is made available and managed with high computational capability.

REFERENCES

Buyya, R., Broberg, J., & Goscinski, A. M. (Eds.). (2010). *Cloud computing: Principles and paradigms*. John Wiley & Sons.

Baun, C., Kunze, M., Nimis, J., & Tai, S. (2011). *Cloud Computing: Web-based dynamic IT services*. Springer Science & Business Media. doi:10.1007/978-3-642-20917-8

Buyya, R., Yeo, C. S., Venugopal, S., Broberg, J., & Brandic, I. (2009). Cloud computing and emerging IT platforms: Vision, hype, and reality for delivering computing as the 5th utility. *Future Generation Computer Systems*, *25*(6), 599–616. doi:10.1016/j.future.2008.12.001

Schatz, M. C., Langmead, B., & Salzberg, S. L. (2010). Cloud computing and the DNA data race. *Nature Biotechnology*, *28*(7), 691–693. doi:10.1038/nbt0710-691 PMID:20622843

Matsunaga, A., Tsugawa, M., & Fortes, J. (2008). Cloudblast: Combining mapreduce and virtualization on distributed resources for bioinformatics applications. *Proceedings of the IEEE Fourth International Conference on eScience*, (pp. 222–229). doi:10.1109/eScience.2008.62

Schatz, M. C. (2009). CloudBurst: Highly sensitive read mapping with MapReduce. *Bioinformatics (Oxford, England)*, *25*(11), 1363–1369. doi:10.1093/bioinformatics/btp236 PMID:19357099

Wall, D. (2010). Cloud computing for comparative genomics. *BMC Bioinformatics*, *11*, 259.

Langmead, B., Schatz, M. C., Lin, J., Pop, M., & Salzberg, S. L. (2009). Searching for SNPs with cloud computing. *Genome Biology*, *10*, R134.

Giardine, B. (2005). Galaxy: A platform for interactive large-scale genome analysis. *Genome Research*, *15*(10), 1451–1455. doi:10.1101/gr.4086505 PMID:16169926

Calabrese, B., & Cannataro, M. (2016). *Cloud Computing in Bioinformatics: Current solutions and challenges.* PeerJ Preprints 4:e2261v1

Shanahan, H. P., Owen, A. M., & Harrison, A. P. (2014). Bioinformatics on the Cloud Computing Platform Azure. *PLoS ONE*, *9*(7), e102642. doi:10.1371/journal.pone.0102642 PMID:25050811

Leite, A. F., & Magalhaes Alves de Melo, A. C. (2012). Executing a biological sequence comparison application on a federated cloud environment. In *2012 19th International Conference on High Performance Computing*. IEEE. doi:10.1109/HiPC.2012.6507500

Sakr, M. F. (2010). *Introduction to cloud computing*. Retrieved from http://www.qatar.cmu.edu/~msakr/15319-s10/lectures/lecture02.pdf

Sen, J. (n.d.). *Security and Privacy Issues in Cloud Computing*. Retrieved from https://arxiv.org/ftp/arxiv/papers/1303/1303.4814.pdf

Firdhous, M., Ghazali, O., & Hassan, S. (2011). A trust computing mechanism for cloud computing. In *Kaleidoscope 2011: The Fully Networked Human? - Innovations for Future Networks and Services (K-2011),Proceedings of ITU* (pp. 1–7).

Afgan, E., Chapman, B., Jadan, M., Franke, V., & Taylor, J. (2012). Using cloud computing infrastructure with CloudBioLinux, CloudMan, and Galaxy. Current protocols in bioinformatics editorial board. doi:10.1002/0471250953.bi1109s38

KEY TERMS AND DEFINITIONS

Bioinformatics: The application of computer, mathematics and statics for solving biological problems.

Cloud Computing: The concept adopted to use a network for a remote access from of server hosted on Internet. The server is used to store, manage, and process data, which prevent local expenditure of the individual or institution.

Distributed Environment: A set of conputing machines placed at different location and connected to each other through a network to form single system.

WSDL: WSDL is an XML format for describing network services as a set of endpoints operating on messages containing either document-oriented or procedure-oriented information.

SOAP (Simple Object Access Protocol): A protocol which allows to transfer web service messages over network using HTTP (Hyper text transfer protocol) and XML.

Chapter 14
Open Access Journal in Bioinformatics:
A Study

Rekha Pareek
University of Kota, India

Sudhir Kumar
Vikram University, India

ABSTRACT

Bioinformatics is rapidly growing, interdisciplinary field of science, where methods from information technology, computer science, mathematics, and statistics are used to solve problems of biological science. To access latest scholarly articles in such an important branch one cannot deny the importance of open access journals. In this chapter an attempt has been made to access the current status of open access journals of bioinformatics which are covered by Directory of Open Access Journals (DOAJ) on various parameters like country and language of publication, their currency, impact factor, article processing charges, copyright licensing model they are using, platform for hosting and their coverage in abstracting/indexing databases.

DOI: 10.4018/978-1-5225-1871-6.ch014

INTRODUCTION

The tremendous growth in internet services and users as well since the 1990s led to universal sharing of knowledge and access to information resources. Similarly, scholarly communication channels also got affected and the internet made them accessible and enhances their readership. Dissemination of scholarly contents through Open access (OA) open new vistas around the globe and these contents are available through the internet in various forms, free of charge and free from copyright and licensing restrictions. Suber (2012) defines OA "*Open Access literature is digital, online, free of charge, and free of most copyright and licensing restrictions*". The OA movement uses the term Gold OA for OA delivered by journals, regardless of the journal's business model, and Green OA for OA delivered by repositories. Self-archiving is the practice of depositing one's own work in an OA repository. All three of these terms were coined by Harnad (2015). Budapest Open Access Initiative in 2002 made revolutionary growth in Open Access contents.

The biggest leap in open access publishing came with the "Budapest Open Access Initiative" (2002), which was aimed to provide free access to refereed articles on the Internet. Budapest Open Access Initiative was launched on 14 February 2002, which is becoming a most popular initiative for scholars and helping them to self-archive their refereed journal articles online. This is further assisting in the establishment of alternative journals that are committed to offering free and unrestricted online access to published articles.

The 2002 Budapest Open Access Initiative's proposed a comprehensive definition of open access as follows:

Free availability on the public internet, permitting any users to read, download, copy, distribute, print, search, or link to the full texts of these articles, crawl them for indexing, pass them as data to software, or use them for any other lawful purpose, without financial, legal, or technical barriers other than those inseparable from gaining access to the internet itself. (Budapest Open Access Initiative, 2002)

There have been various other initiatives, the other definition came from Berlin Declaration on Open Access to Knowledge in the Sciences and Humanities (2003), ACRL Principles and Strategies for the Reform of Scholarly Communication (2003), UN World Summit on the Information Society Declaration of Principles and Plan of Action (2003), OECD Declaration on Access to Research Data from Public Funding (2004), IFLA Statement on Open Access to Scholarly Literature and Research Documentation (2004), and Wellcome Trust Position Statement on Open Access (2005). The initiatives have demonstrated the open access movement. The momentum has continually gained momentum from library and information point of view, funding agencies scholarly, scholarly societies, and institutions of higher education (Zhang 2007). Because open access primarily focuses on research journals, Lynch (2006, p. 5) concludes that "open access to the research journal literature is inevitable, and that open access has. Similar sentiments and beliefs are reflected in the Bethesda Statement (Patrick et. al. 2003) as well. According to Morrison (2015) the size of contents in The Bielefeld Academic Search Engine is over 71 million documents and in Internet archive it is 7.8 million. The Directory of Open Access (DOAJ) Journals is showing consistent strong growth. Over the past year, the growth in articles that can be retrieved through a DOAJ article-level search grew by over a quarter of a million articles for a total of over 1.8 million articles.

Well established publication houses like Elsevier, Taylor, and Francis, Springer and others also introducing open access journals (Rufai et al, 2011). Many of the existing journals of the repute also adopted the open access policy to reach their readers. At present DOAJ listed 11445 journals of various disciplines and this no. is quite higher than 9919 in 2014 as reported by Pujar (2014). Unlike other academic discipline, bioinformatics is in its infancy and still if you go through the website of DOAJ, it accounts 68 journals, search by keyword of the publisher. Though some of them are not exactly related to bioinformatics, some have been merged with other journals of the publication house and some have been discontinued. Websites of these journals visited thoroughly and 37 continued journals were found, strictly related to bioinformatics. In this article, these 37 journals were considered for the study of their language of publication, a platform for hosting, indexing by abstracting journals, currency etc.

BACKGROUND

The Budapest statement defined open access as:

There are many degrees and kinds of wider and easier access to this literature. By 'open access' to this literature, we mean its free availability on the public internet, permitting any users to read, download, copy, distribute, print, search, or link to the full texts of these articles, crawl them for indexing, pass them as data to software, or use them for any other lawful purpose, without financial, legal, or technical barriers other than those inseparable from gaining access to the internet itself. The only constraint on reproduction and distribution and the only role for copyright in this domain should be to give authors control over the integrity of their work and the right to be properly acknowledged and cited. (Budapest Open Access Initiative, 2002)

Tremendous growth and development in numbers of open access journal itself stat its success story. It was 1200 open access journals in 2004 whereas just five in 1992 (Falk, 2004). This number reached to more than eleven thousand journals available only at DOAJ. Lynch (2006) also state open access as an increased elimination of barriers to use scholarly literature by the reader. Nicholas et al (2005) commented on the importance of Open Access Journals as it is possible to "read, download, copy, distribute and print articles and other materials freely". Top journals like Nature, Wall street journal, and Scientist ranked open access among their top stories in 2003 (Willinsky, 2006). McVeigh (2004) also documents that number of open access journals is growing in citation index provided by ISI Tomson. Borgman (2007) also studied open access journals of various fields. Gul et al (2008) also elaborate the growth of open access journals in Scopus citation database.

Bioinformatics is rapidly growing, interdisciplinary field of science that applies methods from information technology, computer science, mathematics and statistics to solve problems in biological science (Molatudi *et al.* 2009). There are growing numbers of massive and heterogeneous biological data sets, including genomics, transcriptomics and proteomics data. Though bioinformatics is a very important discipline but the study of open access journals in the field of Bioinformatics is still lacking. This chapter is an attempt to study the prevalence of bioinformatics journals available through DOAJ.

OBJECTIVES

The objectives of the paper include following:

1. To know the present status of open access journals in Bioinformatics,
2. To know the impact factor of these Journals,
3. To find out the Licensing model,
4. To ascertain the coverage of open access journals by indexing databases, and
5. To know the year of inception of the journals and the year of accepting the open access policy.

SCOPE AND LIMITATIONS

The scope of the study is limited to journals strictly adhere to discipline bioinformatics and only covered in Directory of Open Access Journals (DOAJ). This study does not include journals covered by any other directories, search engines or individual titles available on World Wide Web.

METHODOLOGY

Journals were identified by subject search keyword 'Bioinformatics' in DOAJ. 68 journals were found. Web site of Individual journal was visited to check current status of journal. It was found that some of them were wrongly classified as they were related to computer science or mathematics instead of bioinformatics. Some of the journals were discontinued or merged with other journals of similar discipline from same publication house. It was found that 37 journals out of extracted 68 were strictly adhering to the field of bioinformatics. A spreadsheet was prepared using information related to these 37 journals.

RESULTS AND DISCUSSION

Countrywide Publication

There were 37 open access journals of bioinformatics published from 17 countries. Among these maximum 9 are published from United Kingdom (24.32%), respectively followed by 4 (10.81%) from Egypt and New Zealand, 3 (8.10%) form each Switzerland and Iran, 2 (5.40%) from India and Germany and 1(2.70%) from Singapore, Australia, Ukraine, Greece, Japan, Cuba, Bulgaria, Sweden, US and Malaysia. See Table 1.

Language of Publication

While looking at language of publication of these journals, it was found that 34 (91.89%) journals are being published in English and rest of the journals are being published one of each in Russian, Spanish and Persian. See Table 2.

Continuous and timely publication of a journal is not an easy task, it requires the active participation of subject experts and money as well consistently. When websites of these 37 journals visited individually it was found that all the journals are sustaining their consistency and are regularly being published.

Coverage in Abstracting and Indexing Databases

All the 37 journals have been indexed in any or some of the abstracting/indexing databases such as Google scholar, Scopus, Web of Science, CABS, MAPA and even covered by EBSCO, Proquest like databases which make distribution of contents to the end users. Figure 1 illustrates the indexing of the open access journal in Web of Science.

Table 1. Country wise distribution of journals

Country	No. of Journals	Percentage
United Kingdom	9	24.32
Egypt	4	10.81
New Zealand	4	10.81
Switzerland	3	8.10
Iran	3	8.10
India	2	5.40
Germany	2	5.40
Singapore	1	2.70
Australia	1	2.70
Ukraine	1	2.70
Greece	1	2.70
Japan	1	2.70
Cuba	1	2.70
Bulgaria	1	2.70
Sweden	1	2.70
US	1	2.70
Malaysia	1	2.70

Table 2. Language of publication of journals

Language	No. of Journals	Percentage
English	34	91.89
Russian	1	2.70
Spanish	1	2.70
Persian	1	2.70

The BMC Bioinformatics has highest number of papers published on different topics of bioinformatics, followed by 'Bioinformatics' journal, and Nucleic Acids Research. Again, In terms of number of articles published on bioinformatics by various journals available in open access mode, it is found that the BMC bioinformatics, has highest number of article indexed in Web of Science are available in Open Access Mode. See Figure 2.

Figure 1. Web of science indexing of open access journals in bioinformatics
For a more accurate representation of this figure, please see the electronic version.

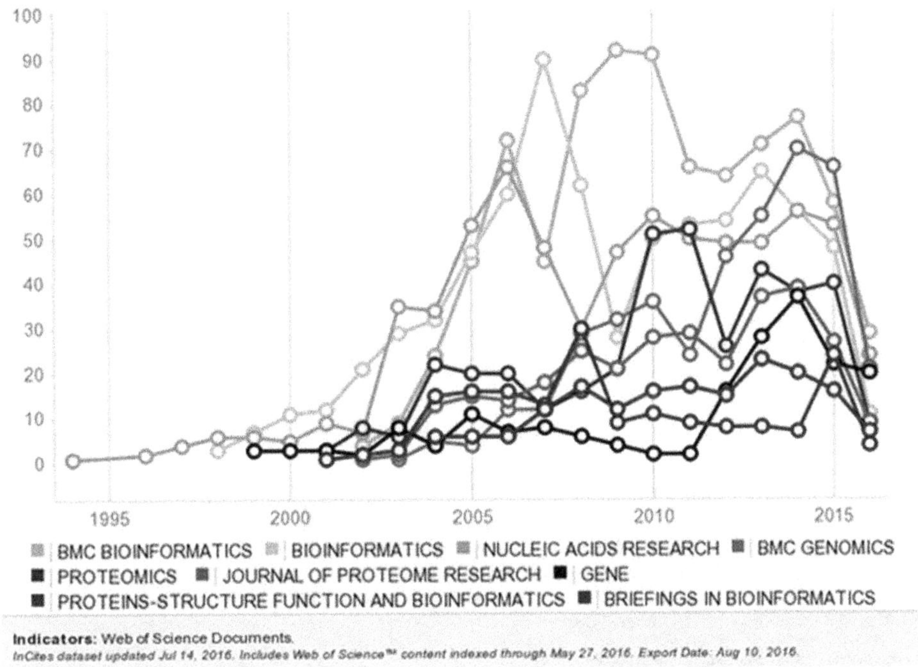

**For a more accurate representation of this figure, please see the electronic version.*

Article Processing Charge

By analyzing individual websites of the journals it was observed that most of the journals charging a fee for article processing/publication. Only two titles are found those are not charging any article processing/publication fee. Since publishing a journal is a costly affair and studies showed that the process of peer review costs on an average 400 USD per article (Rawland, 2002). Thus the open access journals have to charge article processing fee from the author to meet out their expenses.

Impact Factor

The impact factor of a journal is measured by a total number of articles published in last two years and number of citation hence it is an important measurement of quality of the journal. When these 37 journals were analyzed it was found that only 9 journals having impact factor as per the Thomson Reuter's journal citation report. The impact factor ranges from Nucleic Acids Research had impact factor higher than all these journals, but from 2011 onwards, Genome Biology has overpassed all the journals with an impact factor of 14.63 (JCR 2015). Nucleic Acids Research

Figure 2. Availability of open access articles indexed in Web of Science
For a more accurate representation of this figure, please see the electronic version.

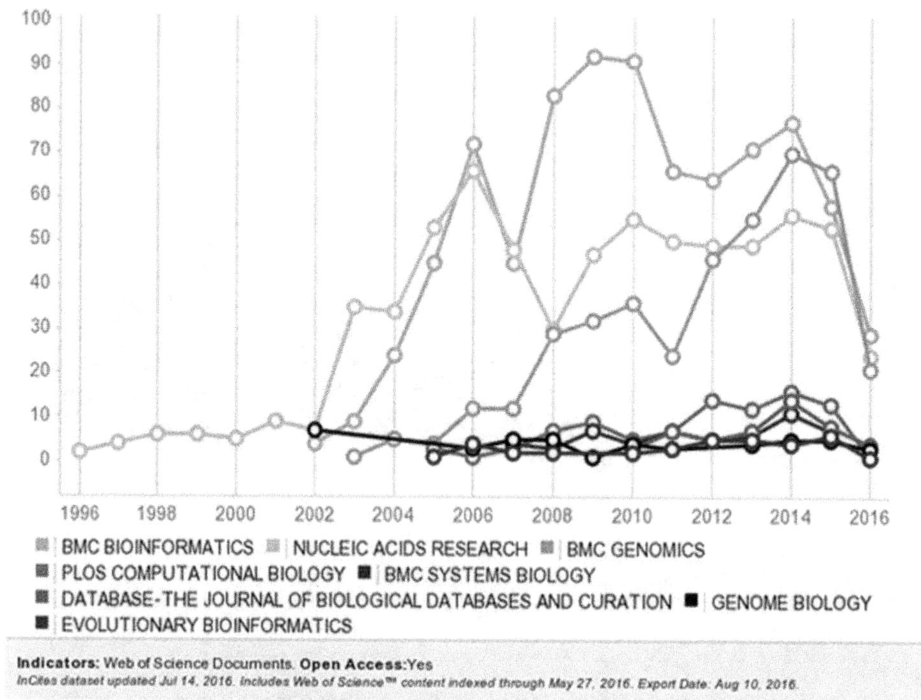

**For a more accurate representation of this figure, please see the electronic version.*

is another journal which has continuously maintained high impact factor of 9.202 (JCR 2015). See these trends in Figure 3.

Licensing Model

Most of the journals of bioinformatics (27, i.e. 72.97%) adopted Creative Commons (CC BY) license and that is the true spirit of open access, that distributing contents and allowed users to remix, tweak, and build upon that matter even commercially without any charge. 6 out of 37 journals (16.21%) journals adopted CC BY-NC license that means they allow users to remix, tweak and build upon the matter non-commercially. The user must acknowledge the author. 3 journals (8.10%) opt CC BY-NC-ND model of licensing that means they only allow the user to download the articles and share them with others but they can change the matter in any way or use it commercially. Only one journal (2.70%) opt the CC-BY-SA license that means allow user to remix, tweak and builds upon the work even for commercially purposes, and even derivative work also carry the similar licensing terms. Some

Figure 3. Trends of impact factor of open access journals in bioinformatics
For a more accurate representation of this figure, please see the electronic version.

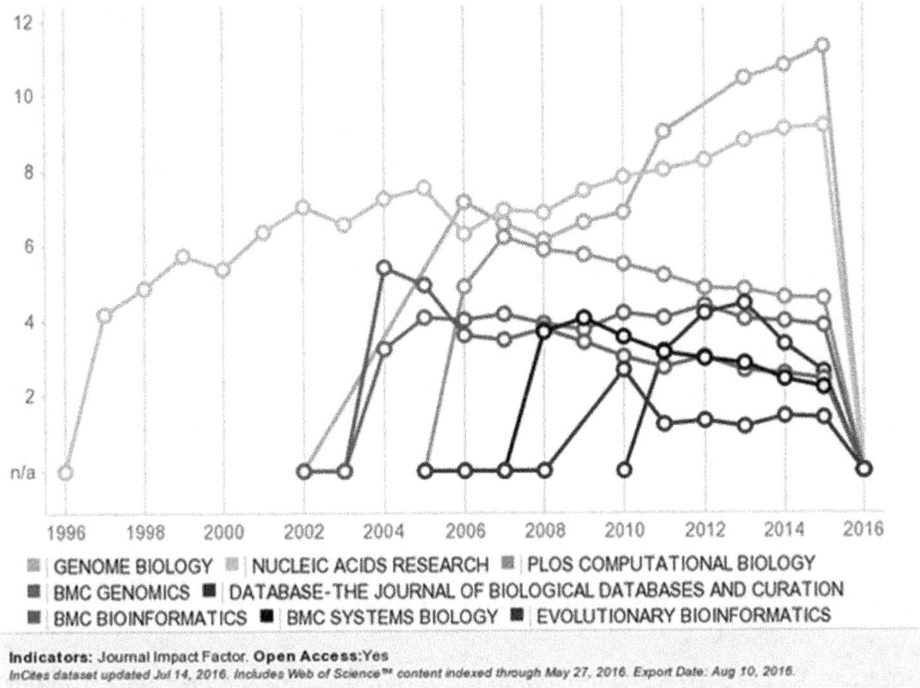

**For a more accurate representation of this figure, please see the electronic version.*

journal (2.70%) opt CC BY-NC-SA license model. It means the user may remix, tweak and build upon the matter for non-commercially.

Year of Inception vs. Open Access

Out of these 37 journals, it was observed that 13 (35.13%) journals provide open access from their inception while remaining 22 (59.45%) adopt open access model in later years. It is a clear cut indication that popularity of open access is gaining tempo day by day and such journals opt this model which were not in open access at the time of their inception. Even some of the well-reputed publication houses like Elsevier, Springer and Taylor and Francis also opt to publish in open access.

CONCLUSION

Being a very new discipline of academics, a number of open access journals in bioinformatics is very less in compare to another field of science. Even some journals are wrongly classified with bioinformatics while they belong to Computers or mathematical sciences. The UK is leading in no. of journals published from any country. If all the 68 journals took into consideration India is the leading publisher, this fact clearly reveals that developing countries still mixing this branch of biological science with physical and mathematical sciences. On the other hand, even other developed countries like Germany, Japan, Australia, and Singapore are also publishing one or two open access journals in bioinformatics. While countries like Egypt, New Zealand, Iran, and Switzerland are leading just behind the UK. Most of the journals are publishing their content in English; the contribution of other languages is negligible. All the journals are covered by abstracting services and use different models of creative common licensing, this makes the contents available to the end users.

REFERENCES

Borgman, C. L. (2007). *Scholarship in the digital age: information, infrastructure, and the internet.* Cambridge, MA: MIT press.

Budapest Open Access Initiative. (2002), available at: www.soros.org/openaccess/read.shtml

Budapest Open Access Initiative (BOAI), (2002). *Interblending & Document Supply, 30*(2).

Falk, H. (2004). Open access gains momentum. *The Electronic Library, 22*(6), 527–530. doi:10.1108/02640470410570848

Gul, S., Vani, Z. A., & Majeed, I. (2008). Open access journals: A global perspective. *Trends in Information Management, 4*(1), 1–19.

Harnad, S. (2015). *Definition of open access.* Retrieved March 14, 2016, from http://openaccess.eprints.org/index.php?/categories/19-Definition-of-Open-Access

Lynch, C. (2006). Improving access to research results: six points. *ARL Bimonthly Report, 248*, 5-7.

Lynch, C. (2006). Improving access to research results: six points. ARL Bimonthly Report, 248, 5-7.

Lynch, C., & Lippincott, J. K. (2005). Institutional repository development in the United States as of early 2005. *D-Lib Magazine, 11*(9). doi:10.1045/september2005-lynch

McVeigh, M. E. (2004). *Open access journals and the ISI citation database: Analysis of impact factor and citation patterns*. Thomson scientific white paper. Retrieved on March 13, 2016, From http://www.thomsonisi.com/media/presentrep/essayspdf/openaccesscitation2.pdf

Molatudi, M., Molotja, N., & Pouris, A. (2009). A bibliometric study of bioinformatics researches in South Africa. *Scientometrics, 81*(1), 477–489. doi:10.1007/s11192-007-2048-6

Morrison, H. (2015). *The dramatic growth of open access*. Retrieved on March 2, 2016, from http://poeticeconomics.blogspot.in

Nicholas, D., Huntington, P., & Rowlands, I. (2005). Open access journal publishing: The views of some of the worlds senior authors. *The Journal of Documentation, 61*(4), 497–519. doi:10.1108/00220410510607499

Patrick, O. B. (2003). *Bethesda Statement on Open Access Publishing*. Retrieved on Feb. 21, 2016 from http://www.earlham.edu/~peters/fos/bethesda.htm

Pujar, S. M. (2014). Open access journals in library and information science: A study. *Annals od Library and information. Studies, 61*, 199–202.

Rowland, F. (2002). The peer review process. *Learned Publishing, 15*(4), 247–258. doi:10.1087/095315102760319206

Rufai, R., Gul, S., & Shah, T. A. (2011). Open access journals in LIS. *Trends in Information Management, 7*(2), 218-228.

Suber, P. (2012). *Open Access*. Cambridge, MA: The MIT Press.

Willinsky, J. (2006). *The access principle- the case for open access to research and scholarship*. Cambridge, MA: The MIT Press.

Zhang, S. L. (2007). The flavors of open access. *OCLC Systems & Services: International Digital Library Perspectives, 23*(3), 229 – 234.

KEY TERMS AND DEFINITIONS

Abstracting/Indexing Database: Collections of multiple sources of publications on one platform. They provide up to abstract information at a glance and such online databases also provide a hyperlink to access full article from there itself.

Bioinformatics: Branch of life science in which large data like sequence information are stored, retrieved and analyzed using IT tools. IT tools are also used to interpret the data and prediction of biological information by using available raw data.

Creative Commons: The copyright licenses and tools that forge a balance inside the traditional "all right reserved" setting of copyright law. Creative commons give everyone a standardized way to grant copyright permissions to their creative work.

DOAJ: Directory of Open Access Journals (DOAJ) is a common database of open access journals. DOAJ not only provide easy access to such journals but also felicitate searching journals and articles without visiting individual websites of the journal.

Gold Open Access: Publication of the final article (Version of Record). The article is made freely available online, often after payment of an article publishing charge (APC).

Green Open Access: Archiving of an article on a website or in a repository. This is often the accepted version of an article, not the final published article.

Impact Factor: A mathematical measurement of a journal's impact derived by calculating a total number of articles published in last two years and the total number of citation received in that time by the articles published in that particular journal.

Open Access Journals: Journals which are publishing their contents online and contents are available to access and downloading for users without any charge.

Open Access: Knowledge resources which are publically available on the net for access and use, free of cost.

Open Source Journal Platform: The domains on them open access journals are being hosted to provide public access.

Repositories: An online database or site hosting research materials (articles, research data, presentations, and so on). This material is usually freely available online for anyone to read or download.

Chapter 15
Web Resources on *Zea mays*: An Overview

Shri Ram
Thapar University, India

ABSTRACT

Zea mays (Z mays), commonly known as corn, is a staple food used worldwide. The research field involving Z. mays has huge potential for agricultural scientists, where the new inventions are being used for the better crop protection. Bioinformatics has revolutionized the research where gene sequencing technology has helped a lot in better agricultural practices through mapping. This chapter proposed to review the research involving Z mays and worldwide resources available on the crop.

INTRODUCTION

Maize, commonly known as corn (Scientific Name *Zea mays*) is a large grain domesticated about ten thousand years ago by indigenous people (Bradle, 1980). Maize is the most widely grown crop throughout the globe. In America, about more than 332 million metric tons maize is grown annually, which is highest among all country (FAO, 2013). USA has been the most productive country in terms of total yield according to the Food and Agricultural Organization (Table 1). It has been noticed that the USA has been the most productive county in terms of the over all production of maize crop since 2009.

DOI: 10.4018/978-1-5225-1871-6.ch015

Table 1. Annual production of corn during last five years

Country	Yield in Metric Tons				
	2009	2010	2011	2012	2013
United States of America	33254860	316164930	313948610	273820066	353699441
Brazil	50719822	55364271	55660235	71072810	80538495
Argentina	13121380	22676920	23799830	21196637	32119211
Ukraine	10486300	11953000	22837900	20961300	30949550
India	16719500	21725800	21760000	22260000	23290000
Mexico	20142816	23301879	17635417	22069254	22663953
France	15288217	13974600	15913300	15614100	15053000
Canada	9561200	11714500	10688700	13060100	14193800
South Africa	12050000	12815000	10360000	11830000	12365000
Russian Federation	3963430	3084350	6962440	8212924	11634943
Italy	7877700	8495940	9752592	8194600	6503200
Germany	4527228	4072900	5184000	4991000	4387300

Source: FAO, 2013.

Maize is one of the staple food being consumed worldwide and being grown because of its ability to grow in diverse climatic conditions. The sugar rich varieties called sweet corn are usually grown more due to its consumption by masses. Some other varieties are cultivated for animal feed along with industrial production such as corn oil, and fermentation and distillation into alcoholic beverages, and as chemical feedstocks.

Maize has a long history of genetic and genomic tool development and is considered one of the most accessible higher plant systems. There has been a rapid growth in agricultural research due to the advancement of modern experimental technology, availability of national and international collaboration and funding opportunities.

BIOINFORMATICS AND CROP IMPROVEMENT

The global availability of modern information and communication technology helps in quick access to resources for fundamental research. Agricultural scientists are grabbing such opportunities for the research and development, and share knowledge with the global community. The establishment such Food and Agricultural Organization (FAO) and International Rice Information System

(IRIS) are very active in providing scholarly support both in terms of resources as well as funding. The international agencies are taking initiative towards the improved genetic improvements and production of disease resistance seeds. Simultaneously, due to the research in producing disease resistant varieties of the crop, the productivity has increased. Some of the initiatives by the international organization are related to the germplasm collection, conservation, testing and production of disease resistance variety in both in-vivo and the in-vitro environment. The free exchange of information by the agencies like International Rice Research Institute through its product like International Crop Information Systems (ICIS) and International Rice Information System provides the foundation for a research that adds value to germplasm conservation, evaluation, utilization for future analysis (Portugal et al. 2007).

The proliferation of information and computer technology in various other subjects as revolutionized the research work. The efficiency of computer technology in data analysis has given a platform for more avenue of research. The origin of 'Bioinformatics/Computational Biology' is one of the reasons where computer science has penetration into biological science. Crop science has been impacted due to the application of technology for sustainable plant productivity to provide sufficient food for the increasing human population. In order to maximize the production, technology application is being sought in combination with biological applications. Nowadays, bioinformatics applications and technologies (such as DNA-based technologies) are widely adopted in agricultural production, biological diversity conservation and crop improvement (Wang et al. 2001).The application of genomics and bioinformatics has accelerated to the crop production in changing climatic conditions (Batley and Edwards 2016).

A huge quantity of information on genomic, phenotypic, gene expression, proteomics, phylogenic tree and another related field is available. The availability of large sum of information needs specialized program for analysis to draw a meaningful conclusion. Using these information different kinds of databases, software programs, and tools are being developed to manage information for crop improvement (Brozynska et al. 2015). The bioinformatics is helping to solve such problem to a great extent. The bioinformatics technologies are being utilized for the crop improvement activities and the field of genomics and bioinformatics rapidly expanding its horizon in other field fields of research, due to continued growth and reducing the cost of DNA sequencing and genotyping (Edwards & Batley 2010; Edwards, Batley, Snowdon 2013). There has been the more recent growth of the application of genomics in the area of agriculture.

AGRICULTURAL RESEARCH INVOLVING MAIZE

Maize is one of the staple food among the cereals. It is a rich source of fiber Corn is a good source of pantothenic acid, phosphorus, niacin, dietary fiber, manganese and vitamin B6. Consumption of corns acts as an antioxidants due to the presence of Antioxidant Phytonutrients. The presence of B-complex vitamins (vitamins B1, B5 and folic acid) and its notable protein content (about 5-6 grams per cup), benefits the control of blood sugar. As funding for agricultural research becomes increasingly scarce in many countries, research administrators have come under heightened pressure to ensure that available resources are used efficiently. Accountability increased interest in research impacts assessment, and peer pressure has motivated a large number of scientists to conduct empirical studies to determine the value of agricultural research having their intended effects. A large volume of literature is being generated through these studies and research program which pave the path for future research. The availability of literature in the form of journal articles, conference deliberations, and patents ignite the scope of research. As far as the growth of literature on maize is a concern, a large volume of literature has been published in last twenty-five years. The growth of literature is visible through Figure 1. It is found that two large indexing and abstracting databases, SCOPUS has indexed about 95227 documents, whereas Web of Science has indexed about 87260 documents. The growth of literature on maize research is found to be progressive and since 2001, the growth is extensive. The new research initiatives have been witnessed through this publication growth, which has helped in the improvement in maize crop.

There has been the establishment of the specialized research center and a program which has helped the international research collaboration. One of the programs started in the area of maize research is the CGIAR Research Program, which has an international collaboration between more than 300 partners from the public and private sectors, national institutions, international research organizations and seed companies. This unique partnership seeks to mobilize global resources for maize research and development to achieve a greater strategic impact on maize-based farming systems in Africa, South Asia and Latin America (http://maize.org/) along with the other partner countries. The research agenda of such program includes Sustainable Intensification of Maize-based Farming Systems, Stress-resilient, and Nutritious Maize, and identifying Agricultural Innovation Systems in Maize, which make easy to figure out how knowledge is produced and used, seeing research as one element in the process of innovation (Maize Technical Annual Report 2014).

Figure 1. Growth of literature on maize since 1990
Source: SCOPUS and Web of Science.

International Maize and Wheat Improvement Center

Another center, the International Maize and Wheat Improvement Center (CIMMYT) are working as a front-liner in the area of agricultural research for development, advocating for small farmers and connecting national agricultural research systems for sharing the knowledge, experience, and resources in the area of maize and wheat (http://www.cimmyt.org/). The activity of CIMMYT is focused on the following:

- The conservation and utilization of maize and wheat genetic resources,
- Developing and promoting improved maize and wheat varieties,
- Testing and sharing sustainable farming systems, and
- Analyzing the impact of its work and researching ways for further improvement

Further, the activities of the CIMMYT extends to global maize program (breeding varieties of the crop), conservation of agricultural program (sustainable cropping practices), socioeconomic program (public policy, efficient use of resources, monitoring of global maize and wheat trends, and the understanding of economic, political and institutional environments), and genetic resource program (identification of genetic traits that are identified as priorities for the eco-regional programs). One of the units GREU works in coordination with Crop Research Informatics Lab (CRIL), Germplasm Bank, the Applied Biotechnology Center (ABC), the Seed inspection and distribution unit, and the Seed Health Lab (Morris, Tripp & Dankyi, 1999; Wikipedia 2016).

Indian Institute of Maize Research

As a constituent of the Indian Council of Agricultural Research (ICAR), the Indian Institute of Maize Research (IIMR) is exclusively mandated for maize research. The IIMR is entrusted with the overall responsibility of research, coordination and management of the multidisciplinary programs on maize improvement at the national level and maintaining linkages with international programs.

WEB RESOURCES ON *ZEA MAYS*

DNA Based Sequence Resources: Genome Sequencing Projects

The complete genome sequencing of an organism started through genome sequencing projects has a great revolution in genomic studies. See some genome sequencing projects previously submitted in Table 2. The sequencing projects for both human and animal has ignited the genomics research and further extended to plants. Initially, the publication and accumulation of nucleotide sequences for model plants only provided fundamental information, however now these base sequences from the fundamentals of research in functional plant genetics. Furthermore, DNA sequence data continues to be central in providing the genomic basis for accelerating molecular level understanding of basic biological mechanisms, and the application of such information to crops. Species-specific nucleotide sequences are now providing information related to phenotypic characters, even when based on genome comparative analyses from the few model plants available (Cogburn et al. 2007; Flicek et al. 2008; Paterson 2008; Tanaka et al. 2008).

Table 2. Genome sequencing projects submitted at NCBI, NLM

Db	Count	Description
Assembly	8	Genome assembly information
BioProject	614	Biological projects providing data to NCBI
BioSample	28,778	Descriptions of biological source materials
Clone	1,145,013	Genomic and cDNA clones
Genome	1	Genome sequencing projects by organism
GSS	2,104,136	Genome survey sequences
Nucleotide	1,026,594	DNA and RNA sequences
Probe	58,076	Sequence-based probes and primers
SNP	60,009,738	Short genetic variations
SRA	7,146	High-throughput DNA and RNA sequence read the archive

Source: NCBI.

Molecular Biology Databases about *Zea mays*

Maize as a global crop and is one of the important grain used for genetic and genomic studies. Under laboratory conditions, *Zea mays* are extensively used, however, the development of novel biological tools and resources to aid in the functional identification of gene sequences is greatly needed. Scientists are working under laboratory conditions and compiling lot of data sets for future analysis. These data sets are being stored in such as ways to give the origin of databases. These databases, shown in Table 3, help in comparison with other plant resources. In maize research areas various such databases have been developed which are being used globally for better crop research.

MaizeGDB

MaizeGDB is a database developed for biological information about Zea mays. MaizeGDB helps in a conglomeration of information on Genomic, genetic, sequence, functional characterization, literature along with the person/organization contact information.

Maize Cell Genomics (MCG) Database

MCG is a collection of maize marker lines for studying native gene expression in specific cell types and sub-cellular compartments. The study is conducted using fluorescent proteins (FPs). A large data set of confocal images generated from

Table 3. Various databases on Zea mays

MaizeGDB	Maize Genetics and Genomics Database (MaizeGDB) is a central repository for maize sequence, stock, phenotype, genotypic and karyotypic variation, and chromosomal mapping data. http://maizegdb.org
ZmGDB	The database ZmGDB displays high quality spliced alignments for EST, cDNA, PUT, and model species proteins (Qunfeng Dong et al 2003) http://plantgdb.org/ZmGDB
CornCyc (*Zea mays*)	large-scale computational predictions of enzyme function http://www.plantcyc.org/databases/corncyc/7.0
PMN	Plant Metabolic Pathway Database http://www.plantcyc.org/
The TIGR Maize Database	TIGR is a member of the Consortium for Maize Genomics
Maize Repeat Database	Maize Repeat Database contained 485 characterized maize repeat sequences from the TIGR Cereal Repeat Database http://maize.jcvi.org/repeat_db.shtml
Zea mays ontology	Zea mays ontology – a database of international terms (Leszek et al 2003)
PPDB is a Plant Proteome Database	Plant Proteome DataBase for *Arabidopsis thaliana* and maize (*Zea mays*) http://ensembl.gramene.org/Zea_mays/Info/Index
Maize Sequence Database	ftp://ftp.gramene.org/pub/gramene/maizesequence.org/
Maize (*Zea mays*) Transcription Factor Database	The Maize (*Zea mays*) Transcription Factor Database is a database collecting 764 predicted maize transcription factors (TFs)
PlantTFDB	Plant Transcription Factor Database http://planttfdb_v2.cbi.edu.cn/index.php?sp=Zm

the maize marker lines has been stored in the database to study the sub-cellular structures, protein localization or interactions under various experimental conditions or mutant backgrounds (Krishnakumar, 2015).

Zea mays Plant Structure Ontology Database

The *Zea mays* Plant Structure Ontology database is designed to standardize the variability of terms used to describe various information related to maize such international botanical terms, references, synonyms, and phylogenetic information. The databases help comparative genomic information to elucidate functional aspects of plant biology and to conduct studies of homology (Coe [n.d.]).

CFRAS-DB

The goal of the Corn Fungal Resistance Associated Sequences Database (CFRAS-DB) (http://agbase.msstate.edu/) to identify genes that are important to aflatoxin resistance. A relational database MySQL was used to design the database with a web-based interface that allows researchers to examine microarray, proteomics, QTL studies, and SNP data.

ZmDB

ZmDB is an integrated database for maize genome research. ZmDB originated in 1999 as the web portal for a large project of maize gene discovery, sequencing and phenotypic analysis using a transperson tagging strategy and expressed sequence tag (EST) sequencing.

MAIZE INFORMATION SYSTEM: MAIZE INFORMATICS

Since maize, wheat and rice are three major sources of food around the globe, and the researcher is looking ahead with best research practices to improve the quality of the crop every year. The scientists and research are working in the area to grow a genetically improved variety of these crops. It is always essential to provide scientists with appropriate tools to manage, use and exploit their field information. The field which deals with the management of information (better known as '*Informatics*') in the area of maize was proposed as IBFieldBook (Lugo-Espinosa et al 2013). To manage information, software tools are being used that allow automatic information recollection in the field, either typed, bar code scanned or acquired through Bluetooth devices. Additionally, information must be shared and made it accessible for easy management and analysis. Some other initiatives such as Maize Field Book (Danga et al 2008), act as one of the primary information systems used by the International Maize and Wheat Improvement Center. The field books help in managing field data of independent maize breeding projects, with specialized functions for inventory stock keeping, seed preparation, and data collection.

Shrestha et al. (2010) developed an ontology-based tool for International Maize Information System (IMIS). IMIS is based on the phenotypic and genotypic data exchange for maize. The information system has been developed in collaboration with the Consultative Group on International Agricultural Research (CGIAR; http://www.cgiar.org/) centers for the management and integration of global information on genetic resources, and germplasm improvement for any crop. Further, with the help of Maize breeding programs at CIMMYT (http://beta.cimmyt.org/) the IMIS tool have the capacity of collecting information in the field with respect to phenotypic, genotypic, and environmental information for their experiments generated worldwide, wet lab, and store it into different relational databases. The IMIS (http://imis.cimmyt.org/confluence/display/IMIS/Crop+Finder) is an implementation of the ICIS, which is a computerized database system for general, integrated management and utilization of genealogy, nomenclature, genetic, phenotypic and characterization data for maize.

The International Maize and Wheat Improvement Centre (CIMMYT) with the collaboration of Bioinformatics scientists at James Hutton Institute facilitating the use of DNA tools and novel bioinformatics to help breed climate-resilient maize variety. Some of the project such as *SAGARPA MasAgro Seeds of Discovery project*, more than 5,000 maize genebank has been created using DNA markers. Some of the software and resources developed by Hutton bioinformaticians are *Germinate* (a generic plant genetic resources database), visualization tools such as *Flapjack*, *Helium* and *CurlyWhirly* are being used to support the research work supported by CIMMYT (http://www.cimmyt.org/donor-partner/sagarpa/). Some other tools used for the mapping of Genetic, Physical, and Informatics Resources for Maize has been playing a key role in the crop improvement (Cone et al. 2002).

DISCUSSION AND CONCLUSION

As a cash crop and a model biological system, maize is of great public interest. More than ninety percent of the major food grains are supplemented by three crops wheat, rice, and maize. Every year new facilities and features are being evolved with the advancement of the genomic science and applications of bioinformatics techniques have facilities to discover the problems of genomics. Further, the problems associated with the crop breeding are being solved with the computational approached. The molecular breeding of maize has taken a great leap and its basic involving physical map, fingerprinting methods, genomic mapping has helped in solving the problem associated with the crop improvement experiments. The field work now has become easier due to the advancement of the data handling tools and the field experiments data are now easily available at fast speed in order to examine it into the lab environment. Advancement into the computational approaches using database management system has helped the large scale data analysis within no time and the results are available at very fast speed. The maize research has also gained speed due to the availability of a variety of resources in the form of molecular biology databases, software tools, and computational facilities. The specialized research center established by government bodies have helped in a better research environment for the scientists. This paper has summarized the availability of the different kind of resources in the area of maize research and hence might be useful to the stockholders in the identification of potential information resources for future research and collaboration.

REFERENCES

Batley, J., & Edwards, D. (2016). The application of genomics and bioinformatics to accelerate crop improvement in a changing climate. *Current Opinion in Plant Biology*, *30*, 78–81. doi:10.1016/j.pbi.2016.02.002 PMID:26926905

Beadle, G. W. (1980). The ancestry of corn. *Scientific American*, *242*(1), 112–119. doi:10.1038/scientificamerican0180-112

Brozynska, M., Furtado, A., & Henry, R. J. (2015). Genomics of crop wild relatives: Expanding the gene pool for crop improvement. *Plant Biotechnology Journal*.

Cogburn, Porter, Duclos, Simon, & Burgess, Zhu, … Burnside. (2007). Functional genomics of the chicken—A model organism. *Poultry Science*, *86*, 2059–2094.

Cone, K. C. (2002). Genetic, Physical, and Informatics Resources for Maize. On the Road to an Integrated Map. *Plant Physiology*, *130*(4), 1598–1605. doi:10.1104/pp.012245 PMID:12481043

Danga, J. (2008). *Integration of the maize field book and the International Maize Information System*. Available at http://agris.fao.org/agris-search/search.do?recordID=PH2009000229

Dong, Q., Roy, L., Freeling, M., & Walbot, V. (2003). ZmDB, an integrated database for maize genome research. *Nucleic Acids Research*, *31*(1), 244–247. doi:10.1093/nar/gkg082 PMID:12519992

Edwards, D., & Batley, J. (2010). Plant genome sequencing: Applications for crop improvement. *Plant Biotechnology Journal*, *8*(1), 2–9. doi:10.1111/j.1467-7652.2009.00459.x PMID:19906089

Edwards, D., Batley, J., & Snowdon, R. J. (2013). Accessing complex crop genomes with next-generation sequencing. *Theoretical and Applied Genetics*, *126*(1), 1–11. doi:10.1007/s00122-012-1964-x PMID:22948437

Harper, L., Gardiner, J., Andorf, C., & Lawrence, C. J. (2016). MaizeGDB: The Maize Genetics and Genomics Database. *Methods in Molecular Biology (Clifton, N.J.)*, *1374*, 187–202. doi:10.1007/978-1-4939-3167-5_9 PMID:26519406

Balachandra, R., Metz, T., Bruskiewich, R., & McLaren, G. (2007). International crop information system for germplasm data management. *Methods in Molecular Biology (Clifton, N.J.)*, *406*, 459–471. PMID:18287707

Krishnakumar, V., Choi, Y., Beck, E., Wu, Q., Luo, A., Sylvester, A., & Chan, A. P. et al. (2015). A maize database resource that captures tissue-specific and subcellular-localized gene expression, via fluorescent tags and confocal imaging (Maize Cell Genomics Database). *Plant & Cell Physiology*, *56*(1), e12. doi:10.1093/pcp/pcu178 PMID:25432973

Leszek, P. (2003). Zea mays ontology - a database of international terms. *Trends in Plant Science*, *8*(11), 517–520. doi:10.1016/j.tplants.2003.09.014 PMID:14607095

Lugo-Espinosa, O. (2013). IBFIELDBOOK: An Integrated Breeding Field Book for Plant Breeding. *Revista Fitotecnia Mexicana*, *36*(3), 201–208.

Morris, M. L., Tripp, R., & Dankyi, A. A. (1999). *Adoption and Impacts of Improved Maize Production Technology: A Case Study of the Ghana Grains Development Project*. Economics Program Paper 99-01. Available at http://impact.cgiar.org/pdf/276.pdf

Shrestha, R., Sanchez, H., Ayala, C., Wenzl, P., & Arnaud, E. (2010). *Ontology-driven International Maize Information System (IMIS) for Phenotypic and Genotypic Data Exchange*. doi:10.1038/npre.2010.5029.1

The CIMMYT. (n.d.). Retrieved from https://en.wikipedia.org/wiki/International_Maize_and_Wheat_Improvement_Center

Wang, L., Hall, J. G., Lu, M., Liu, Q., & Smith, L. M. (2001). A DNA computation readout operation based on structure-specific cleavage. *Nature Biotechnology*, *19*(11), 1053–1059. doi:10.1038/nbt1101-1053 PMID:11689851

KEY TERMS AND DEFINITIONS

Maize Informatics: The field which deals with the management of information (better known as '*Informatics*') in the area of maize.

We Resources: The information sources available in the form of databases, web servers in various bioinformatics disciplines.

***Zea Mays*:** World's third largest consumed globally as staple food.

Compilation of References

Adithya Kumari H,, Mahadevamurthy, M., & Chandrashekara, J. (2013). *Use of social networking sites among students of engineering college libraries in Mysore city, Karnataka: A study*. Retrieved August 20, 2015 from http://www.lsrj.in/ArchiveArticles

Adriaans, P., & Zantinge, D. (1996). *Data Mining*. Harlow, UK: Addison-Wesley Longman.

Afgan, E., Chapman, B., Jadan, M., Franke, V., & Taylor, J. (2012). Using cloud computing infrastructure with CloudBioLinux, CloudMan, and Galaxy. Current protocols in bioinformatics editorial board. doi:10.1002/0471250953.bi1109s38

Agrawal, R., Imielinski, T., & Swami, A. (1993) Mining association rules between sets of items in large databases. *SIGMOD Conference*, (pp. 207-216). doi:10.1145/170035.170072

Aho, A. V. (2012). Computation and Computational thinking. *The Computer Journal*, *55*(7), 833–835. doi:10.1093/comjnl/bxs074

Alexis, L., & Mathews, L. (1999). *Fundamentals of Information Technology*. Leon Press.

Allen, B. M., & Hirshon, A. (1998). Hanging together to avoid hanging separately: Opportunities for academic libraries and consortia. *Information Technology and Libraries*, *17*(1), 36–44.

Alon, U. (2007). Network motifs: Theory and experimental approaches. *Nature Reviews. Genetics*, *8*(6), 450–461. doi:10.1038/nrg2102 PMID:17510665

Alpi, K. (2003). Bioinformatics training by librarians and for librarians: developing the skills needed to support molecular biology and clinical genetics information instruction. *Issues in Science and Technology Librarianship, 37*(Spring). Retrieved from http://www.istl.org/03-spring/article1.html

Alpi, K. (2005). *Bioinformatics training by librarians and for librarians: Developing the skills needed to support molecular biology and clinical genetics information instruction*. Retrieved September 15, 2015 from http://www.istl.org/03-spring/article1.html>

Altairac, S. (2006). Naissance d'une banque de données: Interview du prof. Amos Bairoch. *Protéines à la Une*.

Altman, R. A. (1998). Curriculum in Bioinformatics: The time is ripe. *Bioinformatics (Oxford, England), 14*(7), 549–550. doi:10.1093/bioinformatics/14.7.549 PMID:9841111

Altschul, S. F., Gish, W., Miller, W., Myers, E. W., & Lipman, D. J. (1990). Basic local alignment search tool. *Journal of Molecular Biology, 215*(3), 403–410. doi:10.1016/S0022-2836(05)80360-2 PMID:2231712

Amid, C., Birney, E., Bower, L., Cerdeño-Tárraga, A., Cheng, Y., Cleland, I., & Hunter, C. et al. (2011). Major submissions tool developments at the European Nucleotide Archive. *Nucleic Acids Research*, gkr946. PMID:22080548

Andreeva, A., Howorth, D., Brenner, S. E., Hubbard, T. J., Chothia, C., & Murzin, A. G. (2004). SCOP database in 2004: Refinements integrate structure and sequence family data. *Nucleic Acids Research, 32*(suppl 1), D226–D229. doi:10.1093/nar/gkh039 PMID:14681400

Anil Kumar, N.S. (2014). Information seeking behavior by the research scholars & faculty members: A survey of Kurukshetra University, Kurukshetra in the disciplines of life science. *IOSR Journal of Humanities and Social Sciences, 19*(6), 119-138.

Antezana, E., Blondé, W., Egaña, M., Rutherford, A., Stevens, R., De Baets, B., & Kuiper, M et al.. (2009). BioGateway: A semantic systems biology tool for the life sciences. *BMC Bioinf., 10*(Suppl 10), S11. doi:10.1186/1471-2105-10-S10-S11 PMID:19796395

Apweiler, R., Bairoch, A., & Wu, C. H. (2004). Protein sequence databases. *Current Opinion in Chemical Biology, 8*(1), 76–80. doi:10.1016/j.cbpa.2003.12.004 PMID:15036160

Apweiler, R., Bairoch, A., Wu, C. H., Barker, W. C., Boeckmann, B., Ferro, S., & Magrane, M. et al. (2004). UniProt: The universal protein knowledgebase. *Nucleic Acids Research, 32*(suppl 1), D115–D119. doi:10.1093/nar/gkh131 PMID:14681372

Arndt, T., & Currie, J. P. (2010). Web 2.0 for reference services staff training and communication. *Reference Services Review, 38*(1), 152–157. doi:10.1108/00907321011020789

Asemi, A. (2005), Information searching habits of Internet users: a case study on the Medical Sciences University of Isfahan, Iran. *Webology, 2*(1).

Assenov, Y., Ramirez, F., Schelhorn, S.-E., Lengauer, T., & Albrecht, M. (2008). Computing topological parameters of biological networks. *Bioinformatics (Oxford, England), 24*(2), 282–284. doi:10.1093/bioinformatics/btm554 PMID:18006545

Baeza-Yates, R., & Ribeiro-Neto, B. (1999). *Modern Information Retrieval.* Reading, MA: Addison-Wesley.

Bairoch, A. M. (2000). Serendipity in bioinformatics, the tribulations of a Swiss bioinformatician through exciting times! *Bioinformatics (Oxford, England), 16*(1), 48–64. doi:10.1093/bioinformatics/16.1.48 PMID:10812477

Bairoch, A., & Apweiler, R. (1996). The SWISS-PROT protein sequence data bank and its new supplement TREMBL. *Nucleic Acids Research*, *24*(1), 21–25. doi:10.1093/nar/24.1.21 PMID:8594581

Bakshi, K. (2012, March). Considerations for big data: architecture and approach. In *Proceedings of the IEEE Aerospace Conference*. doi:10.1109/AERO.2012.6187357

Balachandra, R., Metz, T., Bruskiewich, R., & McLaren, G. (2007). International crop information system for germplasm data management. *Methods in Molecular Biology (Clifton, N.J.)*, *406*, 459–471. PMID:18287707

Barabasi, A. L., & Oltvai, Z. N. (2004). Network biology: Understanding the cells functional organization. *Nature Reviews. Genetics*, *5*(2), 101–113. doi:10.1038/nrg1272 PMID:14735121

Barile, L. (2011). Mobile technologies for libraries: A list of mobile applications and resources for development. *College & Research Libraries News*, *72*(4), 222–228.

Baro, E. (2015). Toward a Literature-Driven Definition of Big Data in Healthcare. *BioMed Research International*.

Barrett, F. A. (2010). *An analysis of reference services usage at a regional academic health sciences library*. Retrieved from https://indigo.uic.edu/handle/10027/7618

Barrett, T., Troup, D. B., Wilhite, S. E., Ledoux, P., Rudnev, D., Evangelista, C., & Edgar, R. et al. (2007). NCBI GEO: Mining tens of millions of expression profiles—database and tools update. *Nucleic Acids Research*, *35*(suppl 1), D760–D765. doi:10.1093/nar/gkl887 PMID:17099226

Bartlett, J. C., & Neugebauer, T. (2008), A task-based information retrieval interface to support bioinformatics analysis. *Proceedings of International Symposium on Information Interaction in Context*, (pp. 97-101).

Bateman, A., Coin, L., Durbin, R., Finn, R. D., Hollich, V., Griffiths-Jones, S., & Sonnhammer, E. L. et al. (2004). The Pfam protein families database. *Nucleic Acids Research*, *32*(suppl 1), D138–D141. doi:10.1093/nar/gkh121 PMID:14681378

Batley, J., & Edwards, D. (2016). The application of genomics and bioinformatics to accelerate crop improvement in a changing climate. *Current Opinion in Plant Biology*, *30*, 78–81. doi:10.1016/j.pbi.2016.02.002 PMID:26926905

Baun, C., Kunze, M., Nimis, J., & Tai, S. (2011). *Cloud Computing: Web-based dynamic IT services*. Springer Science & Business Media. doi:10.1007/978-3-642-20917-8

Beadle, G. W. (1980). The ancestry of corn. *Scientific American*, *242*(1), 112–119. doi:10.1038/scientificamerican0180-112

Beautyman, W., & Shenton, A. K. (2009). When does Academic Information need stimulate a school-inspired information want? *Journal of Librarianship and Information Science*, *41*(2), 67–80. doi:10.1177/0961000609102821

Benoît, G. (2006). Bioinformatics. *Annual Review of Information Science & Technology*, *39*(1), 179–218. doi:10.1002/aris.1440390112

Benson, D. A., Karsch-Mizrachi, I., Lipman, D. J., Ostell, J., & Sayers, E. W. (2011). GenBank. *Nucleic Acids Research*, *39*(Database issue), D32–D37. doi:10.1093/nar/gkq1079 PMID:21071399

Benson, G. (2013). Nucleic Acids Research Annual Web Server Issue in 2013 – Editorial. *Nucleic Acids Research*, *41*(W1), W1–W2. doi:10.1093/nar/gkt559

Bentley, D. R. (2006). Whole-genome re-sequencing. *Current Opinion in Genetics & Development*, *16*(6), 545–552. doi:10.1016/j.gde.2006.10.009 PMID:17055251

Bergeron, B. P. (2003). *Bioinformatics Computing*. Prentice Hall Professional.

Berman, H. M., Westbrook, J., Feng, Z., Gilliland, G., Bhat, T., Weissig, H., & Bourne, P. E. et al. (2000). The protein data bank. *Nucleic Acids Research*, *28*(1), 235–242. doi:10.1093/nar/28.1.235 PMID:10592235

Berman, H., Henrick, K., & Nakamura, H. (2003). Announcing the worldwide protein data bank. *Nature Structural & Molecular Biology*, *10*(12), 980–980. doi:10.1038/nsb1203-980 PMID:14634627

Beyer, M., & Laney, D. (2012). *The Importance of 'Big Data': A Definition*. Gartner. Retrieved from https://www.gartner.com/doc/2057415/importance-big-data-definition

Bhatnagar, S. S. V., & Srinivasa, S. (2012). *Big Data Analytics*. Springer.

Birney, E., & Clamp, M. (2004). Biological database design and implementation. *Briefings in Bioinformatics*, *5*(5), 31–38. doi:10.1093/bib/5.1.31 PMID:15153304

Borgman, C. L. (2007). *Scholarship in the digital age: information, infrastructure, and the internet*. Cambridge, MA: MIT press.

Brachman, R., & Anand, T. 1996. The Process ofKnowledge Discovery in Databases: A Human-CenteredApproach. In Advances in Knowledge Discovery and Data Mining, (pp. 37–58). Menlo Park, CA: AAAI Press.

Brown, S. M. (2000). *Bioinformatics: a biologist's guide to computing and the internet*. New York, NY: Eaton Publishing.

Brozynska, M., Furtado, A., & Henry, R. J. (2015). Genomics of crop wild relatives: Expanding the gene pool for crop improvement. *Plant Biotechnology Journal*.

Brunak, S., Danchin, A., Hattori, M., Nakamura, H., Shinozaki, K., Matise, T., & Preus, D. (2002). Nucleotide Sequence Database Policies. *Science*, *298*(5597), 1333. doi:10.1126/science.298.5597.1333b PMID:12436968

Bryant, R., Katz, R., & Lazowska, E. (2008). *Big-Data Computing: Creating revolutionary breakthroughs in commerce, science and society*. Washington, DC: Computing Community Consortium.

Budapest Open Access Initiative (BOAI), (2002). *Interblending & Document Supply, 30*(2).

Budapest Open Access Initiative. (2002), available at: www.soros.org/openaccess/read.shtml

Buehler, L., & Rashidi, H. (2005). Bioinformatics basics: Applications in biological science and medicine. Taylor and Francis Group.

Buneman, P., Davidson, S. B., Hart, K., Overton, C., & Wong, L. (1995). *A data transformation system for biological data sources*. Academic Press.

Bushhousen, E., Norton, H. F., Butson, L. C., Auten, B., Jesano, R., David, D., & Tennant, M. R. (2013). Smartphone uses at a university health science center. *Medical Reference Services Quarterly, 32*(1), 52–72. doi:10.1080/02763869.2013.749134

Buyya, R., Broberg, J., & Goscinski, A. M. (Eds.). (2010). *Cloud computing: Principles and paradigms*. John Wiley & Sons.

Buyya, R., Yeo, C. S., Venugopal, S., Broberg, J., & Brandic, I. (2009). Cloud computing and emerging IT platforms: Vision, hype, and reality for delivering computing as the 5th utility. *Future Generation Computer Systems, 25*(6), 599–616. doi:10.1016/j.future.2008.12.001

Calabrese, B., & Cannataro, M. (2016). *Cloud Computing in Bioinformatics: Current solutions and challenges.* PeerJ Preprints 4:e2261v1

Callinan, J. E. (2005). Information Seeking Behavior of Undergraduate Biology Students. *Library Review, 54*(2), 86–99. doi:10.1108/00242530510583039

Cannata, N., Merelli, E. & Altman, R.B (2005), Time to organize the bioinformatics resourceome. *PLoS Computational Biology, 1*, 531-533.

Case, D. O. (2002). *Looking for Information: A Survey of Research on Information Seeking, Needs, and Behavior*. Amsterdam: Academic Press.

Caspi, R., Altman, T., Billington, R., Dreher, K., Foerster, H., Fulcher, C. A., & Karp, P. D. et al. (2014). The MetaCyc Database of metabolic pathways and enzymes and the BioCyc collection of Pathway/Genome Databases. *Nucleic Acids Research, 42*(1), D459–D471. doi:10.1093/nar/gkt1103 PMID:24225315

Catalano, A. (2013). Patterns of graduate students information seeking behavior: A meta-synthesis of the literature. *The Journal of Documentation, 69*(2), 243–274. doi:10.1108/00220411311300066

Center for Biotechnology Information. (n.d.). Retrieved from https://en.wikipedia.org/wiki/National

Cerami, E. G., Gross, B. E., Demir, E., Rodchenkov, I., Babur, O., Anwar, N., & Sander, C. et al. (2011). Pathway commons, a web resource for biological pathway data. *Nucleic Acids Research, 39*(1), D685–D690. doi:10.1093/nar/gkq1039 PMID:21071392

Chandonia, J.-M., Andreeva, A., Howorth, D., Brenner, S. E., Hubbard, T. J., Chothia, C., & Murzin, A. G. (2007). *Data growth and its impact on the SCOP database: new developments.* Berkeley, CA: Ernest Orlando Lawrence Berkeley National Laboratory.

Che, D., Safran, M., and Peng, Z. (2013). *From Big Data to Big Data Mining: challenges, issues, and opportunities*. Springer.

Chen, M., Mao, S., & Liu, Y. (2014). Big data: A survey. *Mobile Networks and Applications, 19*(2), 171–209. doi:10.1007/s11036-013-0489-0

Chui, M., Miller, A., & Roberts, R. (2009). Six ways to make Web 2.0 work. *The McKinsey Quarterly*.

Chung, S. Y., & Wong, L. (1999). Kleisli: A new tool for data integration in biology. *Trends in Biotechnology, 17*(9), 351–355. doi:10.1016/S0167-7799(99)01342-6 PMID:10461180

Cios, K. J., Pedrycz, W., & Swiniarski, R. (1998). *Data Mining Methods for Knowledge Discovery*. Dordrecht: Kluwer. doi:10.1007/978-1-4615-5589-6

Cleveland, A. D., Hannigan, G. G., Bedard, M., Philbrick, J. L., & Turner, P. M. (2007). *Recruiting the next generation of biomedical sciences librarians: Meeting increasingly complex information needs by building on a biomedical sciences education foundation*. Retrieved from http://www.icml9.org/program/track9/public/documents/Ana%20D-110009.doc

Cochrane, G., Akhtar, R., Aldebert, P., Althorpe, N., Baldwin, A., Bates, K., & Browne, P. et al. (2008). Priorities for nucleotide trace, sequence and annotation data capture at the Ensembl Trace Archive and the EMBL Nucleotide Sequence Database. *Nucleic Acids Research, 36*(suppl 1), D5–D12. doi:10.1093/nar/gkm1018 PMID:18039715

Cocosila, M., & Archer, N. (2010). Adoption of mobile ICT for health promotion: An empirical investigation. *Electronic Markets, 20*(3/4), 241–250. doi:10.1007/s12525-010-0042-y

Codd, E. F. (1970). A relational model of data for large shared data banks. *Communications of the ACM, 13*(6), 377–387. doi:10.1145/362384.362685

Cogburn, Porter, Duclos, Simon, & Burgess, Zhu, … Burnside. (2007). Functional genomics of the chicken—A model organism. *Poultry Science, 86*, 2059–2094.

Collen, M. (1995). *A history of medical informatics in the Unites States: 1950 to 1990*. Bethesda, MD: American Medical Informatics Association.

Collen, M. F. (1986). Origin of Medical Informatics. *The Western Journal of Medicine, 145*, 78–86. PMID:3544507

Cone, K. C. (2002). Genetic, Physical, and Informatics Resources for Maize. On the Road to an Integrated Map. *Plant Physiology, 130*(4), 1598–1605. doi:10.1104/pp.012245 PMID:12481043

Consortium, U. (2010). The universal protein resource (UniProt) in 2010. *Nucleic Acids Research, 38*(suppl 1), D142–D148. doi:10.1093/nar/gkp846 PMID:19843607

Consortium, U. (2011). Ongoing and future developments at the Universal Protein Resource. *Nucleic Acids Research, 39*(suppl 1), D214–D219. doi:10.1093/nar/gkq1020 PMID:21051339

Conte, L. L., Ailey, B., Hubbard, T. J., Brenner, S. E., Murzin, A. G., & Chothia, C. (2000). SCOP: A structural classification of proteins database. *Nucleic Acids Research*, *28*(1), 257–259. doi:10.1093/nar/28.1.257 PMID:10592240

Conte, L. L., Brenner, S. E., Hubbard, T. J., Chothia, C., & Murzin, A. G. (2002). SCOP database in 2002: Refinements accommodate structural genomics. *Nucleic Acids Research*, *30*(1), 264–267. doi:10.1093/nar/30.1.264 PMID:11752311

Conti, E. (2008). *Ten years of service to medical and health librarianship*. Retrieved January 22, 2015 from http://www.mahlap.org/index.php?option=com_content&task=view&id=3&Itemid=20

Cooray, D. M. P. N. S. (2012). Molecular biological databases: Evolutionary history, data modeling, implementation and ethical background. *Sri Lanka. Journal of Biomedical Informatics*, *3*(1), 2–11. doi:10.4038/sljbmi.v3i1.2489

Cornell, M., Paton, N. W., Wu, S., Goble, C. A., Miller, C. J., Kirby, P., & Oliver, S. G. et al. (2003). GIMS – an integrated data storage and analysis environment for genomic and functional data. *Yeast (Chichester, England)*, *15*(15), 1291–1306. doi:10.1002/yea.1047 PMID:14618567

Cox, B., & Jantti, M. (2012). Discovering the Impact of Library Use and Student Performance. *Educause Review*. Retrieved from http://er.educause.edu/articles/2012/7/discovering-the-impact-of-library-use-and-student-performance

Croft, D. (2010). Reactome: A database of reactions, pathways and biological processes. *Nucleic Acids Research*. PMID:21067998

Cunningham, D., Grefsheim, S., Simon, M., & Lansing, P. S. (1991). Biotechnology awareness study, Part 2: Meeting the information needs of biotechnologists. *Bulletin of the Medical Library Association*, *79*(1), 45–52. Retrieved from http://www.ncbi.nlm.nih.gov/pmc/articles/PMC225484/ PMID:1998819

Curtis, K. L., Weller, A. C., & Hurd, J. M. (1997). the Information-seeking behavior of health sciences faculty: The impact of new information technologies. *Bulletin of the Medical Library Association*, *85*(4), 402–410. PMID:9431430

Danga, J. (2008). *Integration of the maize field book and the International Maize Information System*. Available at http://agris.fao.org/agris-search/search.do?recordID=PH2009000229

Dayhoff, M. O. (1965). *Atlas of protein sequence and structure*. Retrieved from http://agris.fao.org/agris-search/search.do?recordID=US201300600070

Day, R., Beck, D. A., Armen, R. S., & Daggett, V. (2003). A consensus view of fold space: Combining SCOP, CATH, and the Dali Domain Dictionary. *Protein Science*, *12*(10), 2150–2160. doi:10.1110/ps.0306803 PMID:14500873

De Castro, E., Sigrist, C. J., Gattiker, A., Bulliard, V., Langendijk-Genevaux, P. S., Gasteiger, E., & Hulo, N. et al. (2006). ScanProsite: Detection of PROSITE signature matches and ProRule-associated functional and structural residues in proteins. *Nucleic Acids Research*, *34*(suppl 2), W362–W365. doi:10.1093/nar/gkl124 PMID:16845026

de Groote, S. L., & Dorsch, J. L. (2001). Online journals: Impact on print journal usage. *Bulletin of the Medical Library Association*, *89*(4), 372–378. PMID:11837259

de la Calle, G., García-Remesal, M., Chiesa, S., de la Iglesia, D., & Maojo, V. (2009). BIRI: A new approach for automatically discovering and indexing available public bioinformatics resources from the literature. *BMC Bioinf.*, *10*(1), 320. doi:10.1186/1471-2105-10-320 PMID:19811635

Dean, J., & Ghemawat, S. (2004). *MapReduce: Simplified data processing on large clusters*. Sixth Symposium on Operating System Design and Implementation, San Francisco, CA. Retrieved from labs.google.com/papers/mapreduce.html

Dennett, D. C. (1991). *Consciousness explained*. Boston, MA: Little, Brown, and Co.

Devadason, F. J., & Lingam, P. P. (1996). A Methodology for the Identification of Information Needs of Users. In *62nd IFLA General Conference - Conference Proceedings*. Beijing, China: International Federation of Library Associations and Institutions. Retrieved August 22, 2009, from http://www.ifla.org/IV/ifla62/62-devf.htm

Devadason, F. J., & Lingam, P. P. (1996). *A methodology for the identification of information needs of users*. Retrieved October 20, 2015 from http://archive.ifla.org/IV/ifla62/62-devf.htm

Dong, Q., Roy, L., Freeling, M., & Walbot, V. (2003). ZmDB, an integrated database for maize genome research. *Nucleic Acids Research*, *31*(1), 244–247. doi:10.1093/nar/gkg082 PMID:12519992

Dorsch, J. L. (2000). Information needs of rural health professionals: A review of the literature. *Bulletin of the Medical Library Association*, *88*(4), 346–354. PMID:11055302

Droid, B. (2010). *DNA Alignment*. Retrieved from https://play.google.com/store/apps/details?id=blink.dna.align

Dublin Core Metadata Initiatives. (n.d.). Retrieved from http://dublincore.org/documents/dcmi-terms/

Durand, P., Medigue, C., Morgat, A., Vandenbrouck, Y., Viari, A., & Rechenmann, F. (2003). Integration of data and methods for genome analysis. *Current Opinion in Drug Discovery & Development*, *6*, 346–352. PMID:12833667

Eckerson, W. (2004). Gauge Your Data Warehousing Maturity. *DM Review*, *11*(14), 34.

Eckerson, W. (2011). *Big Data Analytics: Profiling the Use of Analytical Platforms in User Organizations*. The Data Warehousing Institute.

Edwards, D., & Batley, J. (2010). Plant genome sequencing: Applications for crop improvement. *Plant Biotechnology Journal*, *8*(1), 2–9. doi:10.1111/j.1467-7652.2009.00459.x PMID:19906089

Edwards, D., Batley, J., & Snowdon, R. J. (2013). Accessing complex crop genomes with next-generation sequencing. *Theoretical and Applied Genetics*, *126*(1), 1–11. doi:10.1007/s00122-012-1964-x PMID:22948437

Ekanayake, J. (2010). Twister: a runtime for iterative mapreduce. In *Proceedings of the 19th ACM International Symposium on High Performance Distributed Computing*, (pp. 810-818). doi:10.1145/1851476.1851593

Etzioni, O. (1996). The World-Wide Web: Quagmire or goldmine? *Communications of the ACM*, *39*(11), 65–68. doi:10.1145/240455.240473

Falk, H. (2004). Open access gains momentum. *The Electronic Library*, *22*(6), 527–530. doi:10.1108/02640470410570848

Fayyad, U. M., Piatetsky-Shapiro, G., Smyth, P., & Uthurusamy, R. (Eds.). (1996). *Advances in Knowledge Discovery and Data Mining*. Menlo Park, CA: AAAI/MIT Press.

Fayyad, U., Grinstein, G., & Wierse, A. (2001). *Information Visualization in Data Mining and Knowledge Discovery*. San Francisco: Morgan Kaufmann Publishers.

Fayyad, U., Shapiro, G. P., & Smyth, P. (1996). The KDD process for extracting useful knowledge from volumes of data. *Communications of the ACM*, *39*(11), 27–34. doi:10.1145/240455.240464

Fayyad, U., & Uthurusamy, R. (1996). Data mining and knowledge discovery in databases. *Communications of the ACM*, *39*(11), 24–27. doi:10.1145/240455.240463

Fernandez-Suarez, X. M., & Galperin, M. Y. (2012). The 2013 Nucleic Acids Research Database Issue and the online Molecular Biology Database Collection. *Nucleic Acids Research*, *41*(D1), D1–D7. doi:10.1093/nar/gks1297 PMID:23203983

Finn, R. D., Marshall, M., & Bateman, A. (2005). iPfam: Visualization of protein–protein interactions in PDB at domain and amino acid resolutions. *Bioinformatics (Oxford, England)*, *21*(3), 410–412. doi:10.1093/bioinformatics/bti011 PMID:15353450

Finn, R. D., Mistry, J., Schuster-Böckler, B., Griffiths-Jones, S., Hollich, V., Lassmann, T., & Durbin, R. et al. (2006). Pfam: Clans, web tools and services. *Nucleic Acids Research*, *34*(suppl 1), D247–D251. doi:10.1093/nar/gkj149 PMID:16381856

Fiori, A. (2010). *Extraction of biological knowledge using data mining techniques* (Doctoral dissertation). Available from database and data mining group, Politecnico Di Torino, XXII cycle, 2010 Ph.D. Thesis. Retrieved from http://dbdmg.polito.it/wordpress/wp-content/uploads/2010/12/Thesis_Fiori_Alessandro.pdf

Firdhous, M., Ghazali, O., & Hassan, S. (2011). A trust computing mechanism for cloud computing. In *Kaleidoscope 2011: The Fully Networked Human? - Innovations for Future Networks and Services (K-2011),Proceedings of ITU* (pp. 1–7).

Foster, A. E. (2004). A nonlinear model of information seeking behaviour. *Journal of the American Society for Information Science and Technology*, *55*(3), 228–237. doi:10.1002/asi.10359

Franks, B. (2012). *Taming the Big Data Tidal Wave*. New York: Wiley. doi:10.1002/9781119204275

Fresnido, A. M. B., & Yap, J. M. (2014). Academic library consortia in the Philippines: Hanging in the balance. *Library Management, 35*(1/2), 15–36. doi:10.1108/LM-04-2013-0028

Galperin, M. Y., & Cochrane, G. R. (2011). The 2011 Nucleic Acids Research Database Issue and the online Molecular Biology Database Collection. *Nucleic Acids Research, 39*(Database issue), D1–D6. doi:10.1093/nar/gkq1243 PMID:21177655

Gan, S.K., & Poon, J. (2016). The world of biomedical apps: their uses, limitations, and potential. *Scientific Phone Apps and Mobile Devices, 2*(6).

Ganeshbabu, K. (2015). A Study on Role of Big Data in Bioinformatics. *International Journal of Contemporary Research in Computer Science and Technology, 1*(7), 228.

Garfield, E. (2003). The meaning of the Impact Factor. *International Journal of Clinical and Health Psychology, 3*(2), 363–369.

Garg, N., Pundhir, S., Prakash, A., & Kumar, A. (2008). PCR primer design: DREB genes. *J Comput Sci Syst Biol, 1*, 21-40.

Gartner. (n.d.). *Top Predictions for IT Organizations 2012 and Beyond*. Retrieved from http://gartner.com/itipage.jsp?id=1328I132010

Gautham, N. (2007). *Bioinformatics Databases and Algorithms*. Delhi: Narosa Publishing House Limited.

Geer, R. C. (2006). Broad issues to consider for library involvement in bioinformatics. *Journal of the Medical Library Association: JMLA, 94*(3), 286. Retrieved from http://www.ncbi.nlm.nih.gov/pmc/articles/PMC1525323/ PMID:16888662

Gervasio, D. I. (2014). Redefining Virtual: Leveraging Mobile Librarians for SMS Reference. *International Journal of Digital Library Systems, 4*(2), 44–69. doi:10.4018/IJDLS.2014070104

Giardine, B. (2005). Galaxy: A platform for interactive large-scale genome analysis. *Genome Research, 15*(10), 1451–1455. doi:10.1101/gr.4086505 PMID:16169926

Gilbert, D. (2007). Bioinformatics Introduction. Bioinformatics Research Centre. Retrieved from www.brc.dcs.gla.ac.uk/~drg/.../bioinformaticsHM0607/slides/intro.pdf

Golovin, A., Oldfield, T. J., Tate, J. G., Velankar, S., Barton, G. J., Boutselakis, H., & Dimitropoulos, D. (2004). E-MSD: An integrated data resource for bioinformatics. *Nucleic Acids Research, 32*(Database issue), D211–D216. doi:10.1093/nar/gkh078 PMID:14681397

Goodall, D., & Pattern, D. (2011). Academic library non/low use and undergraduate student achievement: A preliminary report of research in progress. *Library Management, 32*(3), 159–170. doi:10.1108/01435121111112871

Goodman, L. (2003). *Making a genesweep: it's official. In BioIT* (p. 12). World News.

Goto, S., Nishioka, T., & Kanehisa, M. (1999). LIGAND database for enzymes, compounds and reactions. *Nucleic Acids Research*, *27*(1), 377–379. doi:10.1093/nar/27.1.377 PMID:9847234

Greene, A. C. (2015). Adapting bioinformatics curricula for big data. *Briefings in Bioinformatics*, 1–8. PMID:25829469

Gretsheim, S. F., & Rankin, J. A. (2007). Information needs and information seeking in a biomedical research setting: A study of scientists and science administrators. *Journal of the Medical Library Association: JMLA*, *95*(4), 426–434. doi:10.3163/1536-5050.95.4.426 PMID:17971890

Grobelnik, M. (2012). *Big-Dat a Tutorial* [PowerPoint slides]. Jozef Stefan Institute. Retrieved from http://www.slideshare.net/markogrobelnik/big-data-tutorial-marko-grobelnik-25-may-2012

Grossman, D. A. (2009). Information Retrieval: Algorithms and Heuristics (2nd ed.). Springer International Edition.

Gul, S., Vani, Z. A., & Majeed, I. (2008). Open access journals: A global perspective. *Trends in Information Management*, *4*(1), 1–19.

Haas, L. M., Schwartz, P. M., Kodali, P., Kotlar, E., Rice, J. E., & Swope, W. C. (2001). DiscoveryLink: A system for integrated access to life sciences data sources. *IBM Systems Journal*, *40*(2), 489–511. doi:10.1147/sj.402.0489

Hadley, C., & Jones, D. T. (1999). A systematic comparison of protein structure classifications: SCOP, CATH and FSSP. *Structure (London, England)*, *7*(9), 1099–1112. doi:10.1016/S0969-2126(99)80177-4 PMID:10508779

Hadoop, A. (2016). *Hadoop*. Retrieved from http://hadoop.apache.org/

Hamm, G. H., & Cameron, G. N. (1986). The EMBL data library. *Nucleic Acids Research*, *14*(1), 5–9. doi:10.1093/nar/14.1.5 PMID:3945550

Hamosh, A., Scott, A. F., Amberger, J. S., Bocchini, C. A., & McKusick, V. A. (2005). Online Mendelian Inheritance in Man (OMIM), a knowledgebase of human genes and genetic disorders. *Nucleic Acids Research*, *33*(suppl 1), D514–D517. doi:10.1093/nar/gki033 PMID:15608251

Han, J., & Kamber, M. (2001). *Data Mining: Concepts and Techniques*. San Diego, CA: Academic Press.

Hannay, J. E., MacLeod, C., & Singer, J. (2009). How do scientists develop and use scientific software. *ICSE Workshop on Software Engineering for Computational Science and Engineering*, Vancouver, Canada. doi:10.1109/SECSE.2009.5069155

Harnad, S. (2015). *Definition of open access*. Retrieved March 14, 2016, from http://openaccess.eprints.org/index.php?/categories/19-Definition-of-Open-Access

Harper, L., Gardiner, J., Andorf, C., & Lawrence, C. J. (2016). MaizeGDB: The Maize Genetics and Genomics Database. *Methods in Molecular Biology (Clifton, N.J.)*, *1374*, 187–202. doi:10.1007/978-1-4939-3167-5_9 PMID:26519406

Hasegawa, H. (2008). *Genome Databases Current Implementation Practices*. Retrieved in October.

Hashimoto, K., Goto, S., Kawano, S., Aoki-Kinoshita, K. F., Ueda, N., Hamajima, M., & Kanehisa, M. et al. (2006). KEGG as a glycome informatics resource. *Glycobiology*, *16*(5), 63R–70R. doi:10.1093/glycob/cwj010 PMID:16014746

Heger, A., Wilton, C. A., Sivakumar, A., & Holm, L. (2005). ADDA: A domain database with global coverage of the protein universe. *Nucleic Acids Research*, *33*(suppl 1), D188–D191. doi:10.1093/nar/gki096 PMID:15608174

Hendrick, B. (2011). *Internet popular with people seeking health information*. Available at: http://women.webmd.com/news/20110512/Internet-popular-with-people-seeking-health-information

Heng, T. S., Painter, M. W., Elpek, K., Lukacs-Kornek, V., Mauermann, N., Turley, S. J., & Kang, J. et al. (2008). The Immunological Genome Project: Networks of gene expression in immune cells. *Nature Immunology*, *9*(10), 1091–1094. doi:10.1038/ni1008-1091 PMID:18800157

Henry, W. M.. et.al (1980), *Online searching: An introduction*. Butterworths and Company.

Hey, T. (2012). The Fourth Paradigm–Data-Intensive Scientific Discovery. In E-Science and Information Management. Springer. doi:10.1007/978-3-642-33299-9_1

Hiew, H. L., & Bellgard, M. A. (2007). Bioinformatics Reference Model: Towards a Framework for Developing and Organising Bioinformatic Resources. In *International Symposium on Computational Models of Life Sciences*. doi:10.1063/1.2816640

Hoffmann, R., & Valencia, A. (2004). A gene network for navigating the literature. *Nature Genetics*, *36*(7), 664. http://www.ihop-net.org/ doi:10.1038/ng0704-664 PMID:15226743

Hoffman, S., & Ramin, L. (2010). Best practices for librarians embedded in online courses. *Public Services Quarterly*, *6*(2-3), 292–305. doi:10.1080/15228959.2010.497743

Holloway, A., van Laar, R. K., Tothill, R. W., & Bowtell, D. (2002). Options available—from start to finish—for obtaining data from DNA microarrays II. *Nature Genetics*, *32*(Supplement), 481–489. doi:10.1038/ng1030 PMID:12454642

Homan, J. M., & McGowan, J. J. (2002). The Medical Library Association: Promoting new roles for health information professionals. *Journal of the Medical Library Association: JMLA*, *90*(1), 80–85. PMID:11838464

Howe, J., (2006, June). The Rise of Crowdsourcing. *Wired, 14*(6).

Hubbard, T. J., Ailey, B., Brenner, S. E., Murzin, A. G., & Chothia, C. (1999). SCOP: A Structural Classification of Proteins database. *Nucleic Acids Research*, *27*(1), 254–256. doi:10.1093/nar/27.1.254 PMID:9847194

Hucka, M., Finney, A., Sauro, H. M., Bolouri, H., Doyle, J. C., Kitano, H., & Wang, J. (2003). The system biology markup language (SBML): A medium for representation and exchange of biochemical network models. *Bioinformatics (Oxford, England)*, *19*(4), 524–531. doi:10.1093/bioinformatics/btg015 PMID:12611808

Huerta, M. (2000). *NIH working definition of bioinformatics and Computational biology*. Retrieved from http://www.bisti.nih.gov/docs/CompuBioDef.pdf

Hulo, N., Bairoch, A., Bulliard, V., Cerutti, L., Cuche, B. A., De Castro, E., & Sigrist, C. J. et al. (2008). The 20 years of PROSITE. *Nucleic Acids Research*, *36*(suppl 1), D245–D249. doi:10.1093/nar/gkm977 PMID:18003654

Humphreys, K. (1967). The subject specialist and the national and University libraries. *Libri*, *17*(1), 29–41.

Ideker, T., Galitski, T., & Hood, L. (2001). A new approach to decoding life: Systems biology. *Annual Review of Genomics and Human Genetics*, *2*(1), 343–372. doi:10.1146/annurev.genom.2.1.343 PMID:11701654

Inmon, W. H. (1996). The data warehouse and data mining. *Communications of the ACM*, *39*(11), 49–50. doi:10.1145/240455.240470

Islam, M. A. (2012). Information-seeking by print media journalist in Rajshahi, Bangladesh. *International Federation of Library Association and Institution*, *38*(4), 283–288.

Ivie, T., McKay, B., May, F., Mitchell, J., Mortimer, H., & Walker, L. A. (2011). Marketing and promotion of library services using Web 2.0: An annotated mediagraphy. *The Idaho Librarian*, *61*(1). Retrieved from http://works.bepress.com/lizzy_walker/3/

Jalloh, B. (2000). A plan for the establishment of a library network or consortium for Swaziland: Preliminary investigations and formulations. *Library Consortium Management: An International Journal*, *2*(8), 165–176.

James, R. A., Rao, M. M., Chen, E. S., Goodell, M. A., & Shaw, C. A. (2012). The Hematopoietic Expression Viewer: Expanding mobileapps as a scientific tool. *Bioinformatics (Oxford, England)*, *28*(14), 1941–1942. doi:10.1093/bioinformatics/bts279 PMID:22576171

Jensen, L. J., Kuhn, M., Stark, M., Chaffron, S., Creevey, C., Muller, J., & von Mering, C. et al. (2009). STRING 8- a global view on proteins and their functional interactions in 630 organisms. *Nucleic Acids Research*, *37*(Database issue).

Jeong, H. et al.. (2000). The large-scale organization of metabolic networks. *Nature*, *407*(6804), 651–654. doi:10.1038/35036627 PMID:11034217

Jeong, H., Mason, S. P., Barabási, A.-L., & Oltvai, Z. N. (2001). Lethality and centrality in protein networks. *Nature*, *411*(6833), 41–42. doi:10.1038/35075138 PMID:11333967

Jetty, S., & Anbu, K. J. P. (2013). SMS-based content alert system: a case with Bundelkhand University Library, Jhansi. *New Library World*, *114*(1-2), 20 – 31.

Kaltenborn, K. F., & Kuhn, K. (2004). The journal impact factor as a parameter for the evaluation of researchers and research. *Revista Espanola de Enfermedades Digestivas*, *96*(7), 460–476. doi:10.4321/S1130-01082004000700004

Kanehisa, M. (1997). A database for post-genome analysis. *Trends in Genetics*, *9*(13), 375–376. doi:10.1016/S0168-9525(97)01223-7 PMID:9287494

Kanehisa, M. (2013). Chemical and genomic evolution of enzyme-catalyzed reaction networks. *FEBS Letters*, *587*(17), 2731–2737. doi:10.1016/j.febslet.2013.06.026 PMID:23816707

Kanehisa, M., & Goto, S. (2000). KEGG: Kyoto encyclopedia of genes and genomes. *Nucleic Acids Research*, *28*(1), 27–30. doi:10.1093/nar/28.1.27 PMID:10592173

Kanehisa, M., Goto, S., Furumichi, M., Tanabe, M., & Hirakawa, M. (2010). KEGG for representation and analysis of molecular networks involving diseases and drugs. *Nucleic Acids Research*, *38*(Databasesuppl 1), D355–D360. doi:10.1093/nar/gkp896 PMID:19880382

Kanehisa, M., Goto, S., Sato, Y., Kawashima, M., Furumichi, M., & Tanabe, M. (2014). Data, information, knowledge and principle: Back to metabolism in KEGG. *Nucleic Acids Research*, *42*(D1), D199–D205. doi:10.1093/nar/gkt1076 PMID:24214961

Kanehisa, M., Sato, Y., Kawashima, M., Furumichi, M., & Tanabe, M. (2016). KEGG as a reference resource for gene and protein annotation. *Nucleic Acids Research*, *44*(D1), D457–D462. doi:10.1093/nar/gkv1070 PMID:26476454

Kangueane, P. (2009). *Bioinformation Discovery: Data to Knowledge in Biology*. Berlin: Springer Science & Business Media. doi:10.1007/978-1-4419-0519-2

Kari, J., & Savolainen, R. (2003). Towards a contextual model of information seeking on the web. *The New Review of Information Behaviour Research*, *4*(1), 155–175. doi:10.1080/14716310310001631507

Karp, P., Riley, M., Saier, M., Paulsen, I., Collado-Vides, J., Paley, S., … Gama-Castro, S. (n.d.). The EcoCyc database. Nucleic Acids Research, 30, 56–58.

Karsch-Mizrachi, I., & Ouellette, B. F. F. (2001). The Genbank Sequence Database. In *Bioinformatics: A Practical Guide to the Analysis of Genes and Proteins* (pp. 45–63). John Wiley & Sons, Inc.

Kashtan, N., Itzkovitz, S., Milo, R., & Alon, U. (2004). Efficient sampling algorithm for estimating subgraph concentrations and detecting network motifs. *Bioinformatics (Oxford, England)*, *20*(11), 1746–1758. doi:10.1093/bioinformatics/bth163 PMID:15001476

Kashyap, H. (2014). Big Data Analytics in Bioinformatics: A Machine Learning Perspective. *Journal of Latex Class Files*, *13*(9).

Katal, A., Wazid, M., & Goudar, R. H. (2013). Big data: Issues, challenges, tools and good practices. In *Proceedings of the 6th International Conference on Contemporary Computing* (IC3 '13). IEEE.

Keating, S. M., Bornstein, B. J., Finney, A., & Hucka, M. (2006). SBMLToolbox: An SBML toolbox for MATLAB users. *Bioinformatics (Oxford, England)*, *22*(10), 1275–1277. doi:10.1093/bioinformatics/btl111 PMID:16574696

Kerith, P. M. (2011). *Librarians: Masters of the info universe*. Retrieved from http://edition.cnn.com/2011/LIVING/04/12/librarians.masters.of.universe/

Khan, N. (2014). Big Data: Survey, Technologies, Opportunities, and Challenges, Hindawi Publishing Corporation. *The Scientific World Journal*, *2014*, 18.

Khanna, V., & Patel, A. (2015). Mobile Application for Global Sequence Alignmentand BLAST – MobSBlast. *International Journal of Computers and Applications*, *120*(13), 1–5. doi:10.5120/21284-4219

Kim, W., Li, M., Wang, J., & Pan, Y. (2011). Biological network motif detection and evaluation. *BMC Systems Biology*, *5*(Suppl 3), S5. doi:10.1186/1752-0509-5-S3-S5 PMID:22784624

Kitano, H. (2002a). Systems biology: A brief overview. *Science*, *295*(5560), 1662–1664. doi:10.1126/science.1069492 PMID:11872829

Kitano, H. (2002b). Computational systems biology. *Nature*, *420*(6912), 206–210. doi:10.1038/nature01254 PMID:12432404

Klein, S. A. (2014). *DNA Sequence*. Retrieved from https://play.google.com/store/apps/details?id=br.com.samuelklein.dna.phonegap.DNA_Sequence

Kohavi, R., Masand, B., Spilipoulou, M., & Srivastava, J. (2002). Web mining. *Data Mining and Knowledge Discovery*, *6*(1), 5–8. doi:10.1023/A:1013266218887

Konieczny, A. (2010). Experiences as an Embedded Librarian in Online Courses. *Medical Reference Services Quarterly*, *29*(1), 47–57. doi:10.1080/02763860903485084 PMID:20391164

Kopp, J. (1998). Library consortia and information technology: The past, the present, the promise. *Information Technology and Libraries*, *17*(1), 7–12.

Kothari, C. R. (2004). Research Methodology Methods and Techniques (2nd ed.). New Delhi: New Age International Publishers.

Krishnakumar, V., Choi, Y., Beck, E., Wu, Q., Luo, A., Sylvester, A., & Chan, A. P. et al. (2015). A maize database resource that captures tissue-specific and subcellular-localized gene expression, via fluorescent tags and confocal imaging (Maize Cell Genomics Database). *Plant & Cell Physiology*, *56*(1), e12. doi:10.1093/pcp/pcu178 PMID:25432973

Kroski, E. (n.d.). *On the Move with the Mobile Web: Libraries and Mobile Technologies*. Available online at http://eprints.rclis.org/12463/1/mobile_web_ltr.pdf

Kuhlthau Carol, C. (1993). A principle of uncertainty for information seeking. *The Journal of Documentation*, *49*(4), 339–355. doi:10.1108/eb026918

Kuhlthau, C. C. (1991). Inside the search process: Information seeking from the users perspective. *Journal of the American Society for Information Science*, *42*(5), 361–371. doi:10.1002/(SICI)1097-4571(199106)42:5<361::AID-ASI6>3.0.CO;2-#

Kulkarni-Kale, U., Sawant, S., & Chavan, V. (2010). Bioinformatics education in India. *Briefings in Bioinformatics*, *11*(6), 616–625. doi:10.1093/bib/bbq027 PMID:20705754

Kumar, D. (2009). *Information needs of faculty members and research scholars of Chaudhary Charan Singh University: A case study*. Retrieved October 20, 2015 fromhttp://digitalcommons.unl.edu/cgi/viewcontent.cgi?

Kumar, S. (2005). *Bioinformatics Web*. Retrieved November 2015, from http://www.bioinformaticsweb.net/data.html

Kumar, S., Konikoff, C., Van Emden, B., Busick, C., Davis, K. T., Ji, S., & Newfeld, S. J. et al. (2011). FlyExpress: Visual mining of spatiotemporal patterns for genes and publications in Drosophila embryogenesis. *Bioinformatics (Oxford, England)*, *27*(23), 3319–3320. doi:10.1093/bioinformatics/btr567 PMID:21994220

Kuonen, D. (2003). Challenges in Bioinformatics for Statistical Data Miners. *Bulletin of the Swiss Statistical Society*, *46*, 10–17.

Lacroix, Z. (2002). Biological data integration: wrapping data and tools. *Information Technology in Biomedicine, IEEE Transactions on*, *6*(2), 123–128. Retrieved from http://ieeexplore.ieee.org/xpls/abs_all.jsp?arnumber=1006299

Lacroix, Z., & Critchlow, T. (Eds.). (2003). *Bioinformatics: Managing scientific data*. San Francisco: Morgan Kaufmann.

Langmead, B., Schatz, M. C., Lin, J., Pop, M., & Salzberg, S. L. (2009). Searching for SNPs with cloud computing. *Genome Biology*, *10*, R134.

Lee, T. J., Pouliot, Y., Wagner, V., Gupta, P., Stringer-Calvert, D. W. J., Tenenbaum, J. D., & Karp, P. D. (2006). BioWarehouse: A bioinformatics database warehouse toolkit. *BMC Bioinformatics*, *7*(1), 170. doi:10.1186/1471-2105-7-170 PMID:16556315

Leinonen, R., Akhtar, R., Birney, E., Bower, L., Cerdeno-Tarraga, A., Cheng, Y., & Cochrane, G. et al. (2010). European Nucleotide Archive. *Nucleic Acids Research*, *39*(suppl1), D28–D31. doi:10.1093/nar/gkq967 PMID:20972220

Leinonen, R., Akhtar, R., Birney, E., Bower, L., Cerdeno-Tárraga, A., Cheng, Y., & Gibson, R. et al. (2010). The European nucleotide archive. *Nucleic Acids Research*.

Leite, A. F., & Magalhaes Alves de Melo, A. C. (2012). Executing a biological sequence comparison application on a federated cloud environment. In *2012 19th International Conference on High Performance Computing*. IEEE. doi:10.1109/HiPC.2012.6507500

Leszek, P. (2003). Zea mays ontology - a database of international terms. *Trends in Plant Science*, *8*(11), 517–520. doi:10.1016/j.tplants.2003.09.014 PMID:14607095

Liang, H., Xue, Y., & Chase, S. K. (2011). Online health information seeking by people with physical disabilities due to neurological conditions. *International Journal of Medical Informatics*, *80*(11), 745–753. doi:10.1016/j.ijmedinf.2011.08.003 PMID:21917511

Li, M., Chen, Y.-B., & Clintworth, W. A. (2013). Expanding roles in a library-based bioinformatics service program: A case study. *Journal of the Medical Library Association: JMLA*, *101*(4), 303–309. doi:10.3163/1536-5050.101.4.012 PMID:24163602

Lim, H., & Venkatesh, T. V. (2000). Bioinformatics in the pre- and post-genomic eras. *Trends in Biotechnology*, *18*(4), 133–135. doi:10.1016/S0167-7799(99)01409-2 PMID:10809530

Lindberg, D. (2000). Internet access to the National Library of Medicine. *Effective Clinical Practice*, *3*(5), 256. PMID:11185333

Lugo-Espinosa, O. (2013). IBFIELDBOOK: An Integrated Breeding Field Book for Plant Breeding. *Revista Fitotecnia Mexicana*, *36*(3), 201–208.

Lynch, C. (2006). Improving access to research results: six points. *ARL Bimonthly Report*, *248*, 5-7.

Lynch, C. (2006). Improving access to research results: six points. ARL Bimonthly Report, 248, 5-7.

Lynch, C. (1999). Medical libraries, bioinformatics, and networked information: A coming convergence? *Bulletin of the Medical Library Association*, *87*, 408–414. PMID:10550026

Lynch, C., & Lippincott, J. K. (2005). Institutional repository development in the United States as of early 2005. *D-Lib Magazine*, *11*(9). doi:10.1045/september2005-lynch

MacMullen, W. J., & Denn, S. O. (2005). Information problems in molecular biology and bioinformatics. *Journal of the American Society for Information Science and Technology*, *56*(5), 447–456. doi:10.1002/asi.20134

Magana, A. J., Taleyarkhan, M., Alvarado, D. R., Kane, M., Springer, J., & Clase, K. (2014). A survey of scholarly literature describing the field of bioinformatics education and bioinformatics educational research. *CBE Life Sciences Education*, *13*(4), 607–623. doi:10.1187/cbe.13-10-0193 PMID:25452484

Maglott, D., Ostell, J., Pruitt, K. D., & Tatusova, T. (2007). Entrez Gene: Gene-centered information at NCBI. *Nucleic Acids Research*, *35*(suppl 1), D26–D31. doi:10.1093/nar/gkl993 PMID:17148475

Mamanova, L., Coffey, A. J., Scott, C. E., Kozarewa, I., Turner, E. H., Kumar, A., & Turner, D. J. et al. (2010). Target-enrichment strategies for next- generation sequencing. *Nature Methods*, *7*(2), 111–118. doi:10.1038/nmeth.1419 PMID:20111037

Manlunching. (2014). *Information Needs and the Uses of Bioinformatics Users of Select Libraries in India: A Study* (Unpublished doctoral thesis). University of Delhi, New Delhi, India.

Manning, C., Raghavan, P., & Schutze, H. (2008). *Introduction to Information Retrieval*. Cambridge University Press. doi:10.1017/CBO9780511809071

Manyika, J. (2011). *Big data: The next frontier for innovation, competition, and productivity*. McKinsey Global Institute. Retrieved from http://www.mckinsey.com/insights/mgi/research/technology_and_innovation/big_data_the_next_frontier_for_innovation

Maojo, V., Iakovidis, I., Martin-Sanchez, F., Crespo, J., & Kulikowski, C. (2001). Medical Informatics and Bioinformatics: European Efforts to Facilitate Synergy. *Journal of Biomedical Informatics*, *34*(6), 423–427. doi:10.1006/jbin.2002.1042 PMID:12198762

Marchionini, G., & Komlodi, A. (2001). Design of interfaces for information seeking. *Annual Review of Information Science & Technology*, 33.

Martin, V. (2013). Developing a Library Collection in Bioinformatics. In Bioinformatics: Concepts, Methodologies, Tools, and Applications (pp. 130 - 150). Idea Group Inc. (IGI).

Martin, J. V. (1996). Subject specialization in British University libraries: A second survey. *Journal of Librarianship and Information Science*, *28*(3), 159–169. doi:10.1177/096100069602800305

Matsunaga, A., Tsugawa, M., & Fortes, J. (2008). Cloudblast: Combining mapreduce and virtualization on distributed resources for bioinformatics applications. *Proceedings of the IEEE Fourth International Conference on eScience*, (pp. 222–229). doi:10.1109/eScience.2008.62

Mayer-Schönberger, V., & Cukier, K. (2013). *Big Data: A Revolution that Will Transform how We Live, Work, and Think*. New York: Houghton Mifflin Harcourt Publishing Company, Business & Economics.

Maynard, N. A. M. (2012). Modeling information-seeking behavior of graduate students at Kuwait. *The Journal of Documentation*, *68*(4), 430–459. doi:10.1108/00220411211239057

McQuiggan, S., Kosturko, L., McQuiggan, J., & Sabourin, J. (2015). *Mobile Learning A Handbook for Developers, Educators, and Learners*. Willey Publishers.

McVeigh, M. E. (2004). *Open access journals and the ISI citation database: Analysis of impact factor and citation patterns*. Thomson scientific white paper. Retrieved on March 13, 2016, From http://www.thomsonisi.com/media/presentrep/essayspdf/openaccesscitation2.pdf

Mcwilliam, H., Valentin, F., Goujon, M., Li, W., Narayanasamy, M., Martin, J., & Lopez, R. et al. (2009). Web services at the european bioinformatics institute-2009. *Nucleic Acids Research*, *37*(suppl 2), W6–W10. doi:10.1093/nar/gkp302 PMID:19435877

Messersmith, D. J., Benson, D. A., & Geer, R. J. (2006). *A Web-based assessment of bioinformatics end-user support services at US universities*. Retrieved September 10, 2015 from http://www.ncbi.nlm.nih.gov/pmc/articles/PMC1525314/

Meyer, E. E. (1997). The first years of the Protein Data Bank. *Protein Science*, *6*(7), 1591–1597. doi:10.1002/pro.5560060724 PMID:9232661

Mi, M., Wu, W., Qiu, M., Zhang, Y., Wu, L., & Li, J. (2016). Use of Mobile Devices to Access Resources Among Health Professions Students: A Systematic Review. *Medical Reference Services Quarterly*, *35*(1), 64–82. doi:10.1080/02763869.2016.1117290 PMID:26794197

Mitra, S., & Acharya, T. (2003). *Data Mining, Multimedia, soft computing and Bioinformatics.* Hoboken, NJ: John Wiley & Sons, Inc.

Mitra, S., Pal, S. K., & Mitra, P. (2002). Data mining in soft computing framework: A survey. *IEEE Transactions on Neural Networks, 13*(1), 3–14. doi:10.1109/72.977258 PMID:18244404

Mojtaba Sookhtanlo, H. M. (2009). Library information–seeking behavior among Undergraduates students of Agricultural extension and education in Iran. *DESIDOC Journal of Library & Information Technology, 29*(4), 12–20. doi:10.14429/djlit.29.256

Molatudi, M., Molotja, N., & Pouris, A. (2009). A bibliometric study of bioinformatics researches in South Africa. *Scientometrics, 81*(1), 477–489. doi:10.1007/s11192-007-2048-6

Morris, M. L., Tripp, R., & Dankyi, A. A. (1999). *Adoption and Impacts of Improved Maize Production Technology: A Case Study of the Ghana Grains Development Project.* Economics Program Paper 99-01. Available at http://impact.cgiar.org/pdf/276.pdf

Morrison, H. (2015). *The dramatic growth of open access.* Retrieved on March 2, 2016, from http://poeticeconomics.blogspot.in

Morya, V., Dewaker, V., Mecarty, S., & Singh, R. (2010). In silico analysis of metabolic pathways for identification of putative drug targets for Staphylococcus aureus. *J Comput Sci Syst Biol, 3*(3), 62-69.

Mosquera, J., & Sanchez-Pla, A. (2008). Serbgo: Searching for the best go tool. *Nucleic Acids Research, 36*(2), W368–W371. doi:10.1093/nar/gkn256 PMID:18480123

Muto, A., Kotera, M., Tokimatsu, T., Nakagawa, Z., Goto, S., & Kanehisa, M. (2013). Modular architecture of metabolic pathways revealed by conserved sequences of reactions. *Journal of Chemical Information and Modeling, 53*(3), 613–622. doi:10.1021/ci3005379 PMID:23384306

Nagarkar, S. (2011). *Web-based Reference Services to Bioinformaticians: Challenges for librarians.* Retrieved from http://conference.ifla.org/past/2011/111-nagarkar-en.pdf

Needleman, S. B., & Wunsch, C. D. (1970). A general methodapplicable to the search for similarities in the amino acidsequence of two proteins. *Journal of Molecular Biology, 48*(3), 443–453. doi:10.1016/0022-2836(70)90057-4 PMID:5420325

Nekrutenko, A., & Taylor, J. (2012). Next-generation sequencing data interpretation: Enhancing reproducibility and accessibility. *Nature Reviews. Genetics, 13*(9), 667–672. doi:10.1038/nrg3305 PMID:22898652

Nel, J. T. (2001). The information seeking process: Is there a sixth sense? *Mousaion, 19*(2), 23–32.

Nepusz, T., Yu, H., & Paccanaro, A. (2012). Detecting overlapping protein complexes in protein-protein interaction networks. *Nature Methods, 9*(5), 471–472. doi:10.1038/nmeth.1938 PMID:22426491

Nfila, R. B., & Darko-Ampem, K. (2002). Developments in academic library consortia from the 1960s to 2000: A review of literature. *Library Management, 23*(4/5), 203–212. doi:10.1108/01435120210429934

Nguyen, P.-V., Verma, C. S., & Gan, S. K.-E. (2014). DNAApp: A mobile application for sequencing data analysis. *Bioinformataics, 30*(22), 3270–3271. doi:10.1093/bioinformatics/btu525 PMID:25095882

Nicholas, D., Huntington, P., & Rowlands, I. (2005). Open access journal publishing: The views of some of the worlds senior authors. *The Journal of Documentation, 61*(4), 497–519. doi:10.1108/00220410510607499

Nicholas, D., Rowlands, I., Clark, D., Huntington, P., Jamali, H. R., & Ollé, C. (2008). The UK scholarly e-book usage: A landmark survey. *Aslib Proceedings, 60*(4), 311–334. doi:10.1108/00012530810887962

Nicholson, J. A. (2010). The third screen as cultural form in North America. In The Wireless Spectrum: The Politics, Practices, and Poetics of Mobile Media. University of Toronto Press.

Nucleic Acids Research – Database Issue. (n.d.). Available online at https://www.oxfordjournals.org/our_journals/nar/database/cap/

ODonovan, C., Martin, M. J., Gattiker, A., Gasteiger, E., Bairoch, A., & Apweiler, R. (2002). High-quality protein knowledge resource: SWISS-PROT and TrEMBL. *Briefings in Bioinformatics, 3*(3), 275–284. doi:10.1093/bib/3.3.275 PMID:12230036

Orengo, C. A., Michie, A., Jones, S., Jones, D. T., Swindells, M., & Thornton, J. M. (1997). CATH–a hierarchic classification of protein domain structures. *Structure (London, England), 5*(8), 1093–1109. doi:10.1016/S0969-2126(97)00260-8 PMID:9309224

Osterbur, D. L., Alpi, K., Canevari, C., & Corley, P. M. (2006). Vignettes: diverse library staff is offering diverse bioinformatics services. *Journal of the Medical Library Association, 94*(3), E188–91.

Pandian, P. M., Jambhekar, A., & Karisiddappa, C. R. (2002). IIM digital library system: Consortia-based approach. *The Electronic Library, 20*(3), 211–214. doi:10.1108/02640470210432357

Panigrahi, P. P., & Singh, T. R. (2012). Computational analysis for functional and evolutionary aspects of BACE-1 and associated Alzheimers related proteins. *International Journal of Computational Intelligence Studies, 1*(4), 322–332. doi:10.1504/IJCISTUDIES.2012.050355

Panigrahi, P. P., & Singh, T. R. (2013). Computational analysis for Alzheimers disease associated pathways and regulatory patterns using Microarray gene expression and network data reveal association with other diseases. *Journal of Theoretical Biology, 334*, 109–121. doi:10.1016/j.jtbi.2013.06.013 PMID:23811083

Pastorelli, M. (2013). HFSP: size-based scheduling for Hadoop. In*Proceedings of the IEEE International Congress on Big Data (BigData'13)*. IEEE.

Patrick, O. B. (2003). *Bethesda Statement on Open Access Publishing*. Retrieved on Feb. 21, 2016 from http://www.earlham.edu/~peters/fos/bethesda.htm

Payberah, A. H. (2014). *Introduction to Big Data* [PowerPoint slides]. Swedish Institute of Computer Science. Retrieved from https://www.sics.se/~amir/files/download/dic/introduction.pdf

Prosser, D. (2005). The economics of Open Access. In *ICOLC autumn 2005 meeting*. Retrieved from http://www.pfsl.poznan.pl/icolc/1/ICOLC_ Econ.ppt

Protein_Data_Bank. (n.d.). *Protein Data Bank Wiki*. Retrieved from https://en.wikipedia.org/wiki/Protein_Data_Bank

Pruitt, K. D., Tatusova, T., & Maglott, D. R. (2005). NCBI Reference Sequence (RefSeq): A curated non-redundant sequence database of genomes, transcripts and proteins. *Nucleic Acids Research*, *33*(suppl 1), D501–D504. doi:10.1093/nar/gki025 PMID:15608248

Pujar, S. M. (2014). Open access journals in library and information science: A study. *Annals od Library and information. Studies*, *61*, 199–202.

Quinn, G. B., Bi, C., Christie, C. H., Pang, K., Prli, A., Nakane, T., & Rose, P. W. et al. (2015). RCSB PDB Mobile: iOS and Android mobile apps to provide data access and visualization to the RCSB Protein Data Bank. *Bioinformatics (Oxford, England)*, *31*(1), 126–127. doi:10.1093/bioinformatics/btu596 PMID:25183487

Rabindra, K., & Maharana, A. P. (2013). Exploring the information seeking behavior of Medical Science. *Information Age*, *7*(1), 18–22.

Raghupathi, W., & Kesh, S. (2007). Interoperable electronic health records design: Towards a service-oriented architecture. *e-Service Journal*, *5*(3), 39–57. doi:10.2979/ESJ.2007.5.3.39

Raghupathi, W., & Raghupathi, V. (2014). Big data analytics in healthcare: Promise and potential. *Health Information Science and Systems*, *2*(1), 3. doi:10.1186/2047-2501-2-3 PMID:25825667

Ram, S., & Rao, N. L. (2012). iBIRA – integrated bioinformatics information resource access: Organizing the bioinformatics resourceome. *Reference Services Review*, *40*(2), 326 – 343.

Ram, S., Anbu, J. P. K., & Kataria, S. (2014). Mobile Information Literacy for libraries: A case study on requirements for an effective Information Literacy Program. In 5-M Libraries: From Devices to People. Facet Publishing.

Ram, S., Sureka, R. K., Sharma, P., & Rao, N. L. (2010). Integrating bioinformatics information resources: Management of information through clustering. In *IEEE Students'Technology Symposium* (TechSym).

Ram, S., Anbu, J. P. K., & Kataria, S. (2011). Responding to users expectation in the library: innovative Web 2.0 applications at JUIT Library: A case study. *Program: Electronic Library and Information Systems*, *45*(4), 452–469. doi:10.1108/00330331111182120

Ram, S., & Rao, N. L. (2012). *i*BIRA – integrated bioinformatics information resource access: Organizing the bioinformatics resourceome. *Reference Services Review*, *40*(2), 326–343. doi:10.1108/00907321211228354

Rao, V., Das, S., & Umari, E. (2009). Glycomics Data Mining. *J Comput Sci Syst Biol*, *2*, 262–265.

Rastogi, S. C., Mendiratta, N., & Rastogi, P. (2006). *Bioinformatics: Methods and Applications – Genomics, Proteomics and Drug Discovery*. New Delhi: Prentice Hall of India.

Reddy, G. S., Srinivasu, R., Rao, M. P. C., & Rikkula, S. R. (2010). Data Warehousing, Data Mining, OLAP and OLTP Technologies are Essential Elements to Support Decision-Making Process inIndustries. *International Journal on Computer Science and Engineering*, *2*(9), 2865–2873.

Reeve, L. H., Han, H., & Brooks, A. D. (2007). The use of domain-specific concepts in biomedical text summarization. *Information Processing and Management: An International Journal*, *43*(6).

Rein, D. C. (2006). Developing library bioinformatics services in context: The Purdue University Libraries bioinformationist program. *Journal of the Medical Library Association: JMLA*, *94*(3), 314. Retrieved from http://www.ncbi.nlm.nih.gov/pmc/articles/PMC1525331/ PMID:16888666

Reitz Joan, M. (Ed.). (2016). *Online Dictionary for Library and Information Science*. Retrieved January 20, 2016 from http://www.abc-clio.com/ODLIS/odlis_c.aspx

Renwick, S. (2005). Knowledge and use of electronic information resources by medical sciences faculty at The University of the West Indies. *Journal of the Medical Library Association: JMLA*, *93*(1), 21–31. PMID:15685270

Rigden, D. J., Fernandez, M., & Galperin, Y. (2016). The 2016 database issue of Nucleic Acids Research and an updated molecular biology database collection. *Nucleic Acids Research*, *44*(D1), D1–D5. doi:10.1093/nar/gkv1356 PMID:26740669

Robson, A., & Robinson, L. (2013). Building on models of information behaviour: Linking information seeking and communication. *The Journal of Documentation*, *69*(2), 169–193. doi:10.1108/00220411311300039

Rojahn, S. Y. (2012, May). *Breaking the Genome Bottleneck*. MIT Technology.

Rother, K., Potrzebowski, W., Puton, T., Rother, M., Wywial, E., & Bujnicki, J. M. (2012). A toolbox for developing bioinformatics software. *Briefings in Bioinformatics*, *13*(2), 244–257. doi:10.1093/bib/bbr035 PMID:21803787

Rowland, F. (2002). The peer review process. *Learned Publishing*, *15*(4), 247–258. doi:10.1087/095315102760319206

Rufai, R., Gul, S., & Shah, T. A. (2011). Open access journals in LIS. *Trends in Information Management, 7*(2), 218-228.

Sagiroglu, S., & Sinanc, D. (2013, May). Big data: a review. In *Proceedings of the International Conference on Collaboration Technologies and Systems (CTS '13)*. IEEE.

Sahoo, G. C., Rani, M., Dikhit, M. R., Ansari, W. A., & Das, P. (2009). Structural modeling, evolution and ligand interaction of KMP11 protein of different leishmania strains. *J Comput Sci Syst Biol*, *2*, 147–158.

Sahu, H. K. & Singh, S. N. (2013). Information seeking behavior of astronomy /astrophysics scientists. *ASLIB Proceedings: New Information Perspectives*, *65*(2), 109-142.

Sakr, M. F. (2010). *Introduction to cloud computing*. Retrieved from http://www.qatar.cmu.edu/~msakr/15319-s10/lectures/lecture02.pdf

Sawyer, S., & Tapia, A. (2005). The sociotechnical nature of mobile computing work: Evidence from a study of policing in the United States. *International Journal of Technology and Human Interaction*, *1*(3), 1–14. doi:10.4018/jthi.2005070101

Schaefer, C., Meier, A., Rost, B., & Bromberg, Y. (2012). SNPdbe: Constructing an nsSNP functional impacts database. *Bioinformatics (Oxford, England)*, *28*(4), 601–602. doi:10.1093/bioinformatics/btr705 PMID:22210871

Schäffer, A. A., Aravind, L., Madden, T. L., Shavirin, S., Spouge, J. L., Wolf, Y. I., & Altschul, S. F. et al. (2001). Improving the accuracy of PSI-BLAST protein database searches with composition-based statistics and other refinements. *Nucleic Acids Research*, *29*(14), 2994–3005. doi:10.1093/nar/29.14.2994 PMID:11452024

Schaff, J., Slepchenko, B., & Loew, L. (2000). Physiological modeling with virtual cell framework. *Methods in Enzymology*, *321*, 1–23. doi:10.1016/S0076-6879(00)21184-1 PMID:10909048

Schatz, M. C. (2009). CloudBurst: Highly sensitive read mapping with MapReduce. *Bioinformatics (Oxford, England)*, *25*(11), 1363–1369. doi:10.1093/bioinformatics/btp236 PMID:19357099

Schatz, M. C., Langmead, B., & Salzberg, S. L. (2010). Cloud computing and the DNA data race. *Nature Biotechnology*, *28*(7), 691–693. doi:10.1038/nbt0710-691 PMID:20622843

Schönbach, C., Kowalski-Saunders, P., & Brusic, V. (2000). Data warehousing in molecular biology. *Briefings in Bioinformatics*, *1*(2), 190–198. doi:10.1093/bib/1.2.190 PMID:11465030

Schreiber, F., & Schwobbermeyer, H. (2005). MAVisto: A tool for the exploration of network motifs. *Bioinformatics (Oxford, England)*, *21*(17), 3572–3574. doi:10.1093/bioinformatics/bti556 PMID:16020473

Schuler, G. D. (1997). Pieces of the puzzle: Expressed sequence tags and the catalog of human genes. *Journal of Molecular Medicine*, *75*(10), 694–698. doi:10.1007/s001090050155 PMID:9382993

Schuler, G. D., Epstein, J. A., Ohkawa, H., & Kans, J. A. (1996). Entrez: Molecular biology database and retrieval system. *Methods in Enzymology*, *266*, 141–162. doi:10.1016/S0076-6879(96)66012-1 PMID:8743683

Sehgal, M., Gupta, R., Moussa, A., & Singh, T. R. (2015). An Integrative Approach for Mapping Differentially Expressed Genes and Network Components Using Novel Parameters to Elucidate Key Regulatory Genes in Colorectal Cancer. *PLoS ONE, 10*(7), e0133901. doi:10.1371/journal.pone.0133901 PMID:26222778

Sen, J. (n.d.). *Security and Privacy Issues in Cloud Computing*. Retrieved from https://arxiv.org/ftp/arxiv/papers/1303/1303.4814.pdf

Shachak, A., Shuval, K., & Fine, S. (2007). Barriers and enablers to the acceptance of bioinformatics tools: A qualitative study. *Journal of the Medical Library Association: JMLA, 95*(4), 454–458. doi:10.3163/1536-5050.95.4.454 PMID:17971896

Shanahan, H. P., Owen, A. M., & Harrison, A. P. (2014). Bioinformatics on the Cloud Computing Platform Azure. *PLoS ONE, 9*(7), e102642. doi:10.1371/journal.pone.0102642 PMID:25050811

Shaw, J. (2014, March-April). *Why "Big Data" Is a Big Deal*. Retrieved from http://harvardmagazine.com/2014/03/why-big-data-is-a-big-deal

Shera, J. H. (1972). *The Foundations of Education for Librarianship*. New York: Becker and Hayes.

Shipman, J. P., Barbara Watstein, S., & Tennant, M. R. (2005). Bioinformatics Librarian: meeting the information needs of genetics and bioinformatics researchers. *Reference Services Review, 33*(1), 12–19. doi:10.1108/00907320410519333

Shrestha, R., Sanchez, H., Ayala, C., Wenzl, P., & Arnaud, E. (2010). *Ontology-driven International Maize Information System (IMIS) for Phenotypic and Genotypic Data Exchange*. doi:10.1038/npre.2010.5029.1

Sigrist, C. J., De Castro, E., Langendijk-Genevaux, P. S., Le Saux, V., Bairoch, A., & Hulo, N. (2005). ProRule: A new database containing functional and structural information on PROSITE profiles. *Bioinformatics (Oxford, England), 21*(21), 4060–4066. doi:10.1093/bioinformatics/bti614 PMID:16091411

Sillitoe, I., Lewis, T. E., Cuff, A., Das, S., Ashford, P., Dawson, N. L., & Lees, J. G. et al. (2015). CATH: Comprehensive structural and functional annotations for genome sequences. *Nucleic Acids Research, 43*(D1), D376–D381. doi:10.1093/nar/gku947 PMID:25348408

Sim, J.Z., Nguyen, P.V., Lim, P.H.J., Su, T.T.C., & Gan, S.K.E. (2015). The Rise of the Mobile Lab: the Use of Smartphone Apps for Biomedical Research. *Asia Pacific Biotech News, 19*(58).

Singh, S., Gupta, S., Nischal, A., Khattri, S., & Nath, R. (2010). Comparative Modeling Study of the 3-D Structure of Small Delta Antigen Protein of Hepatitis Delta Virus. *J Comput Sci Syst Biol, 3*, 1-4.

Singh, K. S. (2007). Information seeking behavior of Agricultural scientists with particular references to their information seeking strategies. *Annals of Library and Information Studies, 54*, 213–220.

Singh, S., & Singh, S. (2004). Need for joining library consortia: a study of Vikram University Library. In *Proceedings of National Seminar on Library Consortia*.

Smith, I. (2005). Achieving readiness for organisational change. *Library Management*, *26*(6/7), 408–412. doi:10.1108/01435120510623764

Smoot, M. E., Ono, K., Ruscheinski, J., Wang, P. L., & Ideker, T. (2011). Cytoscape 2.8: New features for data integration and network visualization. *Bioinformatics (Oxford, England)*, *27*(3), 431–432. doi:10.1093/bioinformatics/btq675 PMID:21149340

Stellrecht, E., & Chiarella, D. (2015). Targeted Evolution of Embedded Librarian Services: Providing Mobile Reference and Instruction Services Using iPads. *Medical Reference Services Quarterly*, *34*(4), 397–406. doi:10.1080/02763869.2015.1082372

Stevens, R., Baker, P., Bechhofer, S., Ng, G., Jacoby, A., Paton, N. W., & Brass, A. et al. (2000). TAMBIS: Transparent access to multiple bioinformatics information sources. *Bioinformatics (Oxford, England)*, *16*(2), 184–185. doi:10.1093/bioinformatics/16.2.184 PMID:10842744

Stoesser, G., Baker, W., van den Broek, A., Garcia-Pastor, M., Kanz, C., Kulikova, T., & Lopez, R. et al. (2003). The EMBL nucleotide sequence database: Major new developments. *Nucleic Acids Research*, *31*(1), 17–22. doi:10.1093/nar/gkg021 PMID:12519939

Stone, G., Ramsden, B., & Pattern, D. (2011). Looking for the link between library usage and student attainment. *Ariadne*, *67*. Retrieved from http://www.ariadne.ac.uk/issue67/stone-et-al/

Suber, P. (2012). *Open Access*. Cambridge, MA: The MIT Press.

Subramanian, R., Muthurajan, R., & Ayyanar, M. (2008). Comparative Modeling an d Analysis of 3-D Structure of EMV2, a Late Embryogenesis Abundant Protein o f Vigna Radiata (Wilczek). *J Proteomics Bioinform*, *1*(8), 401–407. doi:10.4172/jpb.1000049

Sujatha, S. (2014). Exploring information seeking behaviour in the changing ICT environment: A snapshot of Kakatiya University facility. *International Journal of Information Services and Technology*, *1*(1), 11–14.

Suzek, B. E., Huang, H., McGarvey, P., Mazumder, R., & Wu, C. H. (2007). UniRef: Comprehensive and non-redundant UniProt reference clusters. *Bioinformatics (Oxford, England)*, *23*(10), 1282–1288. doi:10.1093/bioinformatics/btm098 PMID:17379688

Tagari, M., Tate, J., Swaminathan, G. J., Newman, R., Naim, A., Vranken, W., & Velankar, S. (2006). E-MSD: Improving data deposition and structure quality. *Nucleic Acids Research*, *34*(90001), D287–D290. doi:10.1093/nar/gkj163 PMID:16381867

Talawar, G. G. M. V. G. (2009). Library consortia in developing countries: An overview. *Program*, *43*(1), 94 – 104. DOI:10.1108/00330330910934138

Tao, D., Demiris, G., Graves, R. S., & Sievert, M. (2003). Transition from in library use of resources to outside library use: the impact of the Internet on information seeking behavior of medical students and faculty.*AMIA Annual Symposium Proceedings*.

Taole, N., & Dick, A. L. (2009). Implementing a common library system for the Lesotho Library Consortium. *The Electronic Library*, *27*(1), 5–19. doi:10.1108/02640470910934551

Tatusova, T. A., Karsch-Mizrachi, I., & Ostell, J. A. (1999). Complete genomes in WWW Entrez: Data representation and analysis. *Bioinformatics (Oxford, England)*, *15*(7), 536–543. doi:10.1093/bioinformatics/15.7.536 PMID:10487861

Thanuskodi, S. (2012). Use of Internet and Electronic Resources among Medical Professionals with special reference to Tamil Nadu: A Case Study. *SRELS Journal of Information Management*, *49*(3), 281–292.

The CIMMYT. (n.d.). Retrieved from https://en.wikipedia.org/wiki/International_Maize_and_Wheat_Improvement_Center

Tomita, M., Hashimoto, K., Takahashi, K., Shimizu, T. S., Matsuzaki, Y., Miyoshi, F., & Hutchison, C. A. et al. (1999). E-cell: A software environment for while-call simulation. *Bioinformatics (Oxford, England)*, *15*(1), 72–84. doi:10.1093/bioinformatics/15.1.72 PMID:10068694

Toomula, N. (2011). Biological databases-integration of life science data. *Journal of Computer Science & Systems Biology*.

Torlone, R. (2015). *Big data: an introduction* [PowerPoint slides]. Università Roma Tre. Retrieved from http://www.dia.uniroma3.it/~torlone/bigdata/L1-Introduzione.pdf

Torres, J. (2014). *Big Data Challenges in Bioinformatics* [PowerPoint Presentation]. Barcelona, Supercomputing Center Computer Science, Department Autonomic Systems and eBusiness Platforms.

Torshin, I. Y. (2006). *Bioinformatics in the Post-genomic Era: The Role of Biophysics*. New York: Nova Publishers.

Turenne, N. (2009). Data mining, a tool for systems biology or a systems biology tool. *J Comput Sci Syst Biol*, *2*(4), 216–218. doi:10.4172/jcsb.1000034e

Vakkari, P. (1998), Task complexity, information types, search strategies and relevance: integrating studies on information seeking and retrieval. In *Exploring the Contexts of Information Behaviour,Proceedings of the Second International Conference on Research in Information Needs, Seeking and Use in Different Contexts*.

Varambhia, H. N. (2013). *Oh BLAST It!* Retrieved from https://play.google.com/store/apps/details?id=com.bioinformaticsapp

Varsale, A., Wadnerkar, A., Mandage, R., & Jadhavrao, P. (2010). Cheminformatics. *J Proteomics Bioinform*, *3*, 253–259. doi:10.4172/jpb.1000148

Velankar, S., Alhroub, Y., Alili, A., Best, C., Boutselakis, H. C., Caboche, S., & Kleywegt, G. J. et al. (2011b). PDBe: Protein Data Bank in Europe. *Nucleic Acids Research*, *39*(Database issue), D402–D410. doi:10.1093/nar/gkq985 PMID:21045060

Vittal, R. S. (2005). Bioinformatics a modern approach. Prentice Hall of India.

Voelkerding, K. V., Dames, S. A., & Durtschi, J. D. (2009). Next-generation sequencing: From basic research to diagnostics. *Clinical Chemistry*, *55*(4), 641–658. doi:10.1373/clinchem.2008.112789 PMID:19246620

Wall, D. (2010). Cloud computing for comparative genomics. *BMC Bioinformatics*, *11*, 259.

Wang, B., Zou, X., & Zhu, J. (2000). Data Assimilation and its Applications. *Proceedings of the National Academy of Sciences of the United States of America*, *97*(21), 11143–11144. doi:10.1073/pnas.97.21.11143 PMID:11027322

Wang, D. (2011). An efficient cloud storage model for heterogeneous cloud infrastructures. *Procedia Engineering*, *23*, 510–515. doi:10.1016/j.proeng.2011.11.2539

Wang, J. T. L., Zaki, M. J., Toivonen, H. T. T., & Shasha, D. E. (2005). *Data Mining in Bioinformatics*. London, UK: Springer-Verlag.

Wang, L., Hall, J. G., Lu, M., Liu, Q., & Smith, L. M. (2001). A DNA computation readout operation based on structure-specific cleavage. *Nature Biotechnology*, *19*(11), 1053–1059. doi:10.1038/nbt1101-1053 PMID:11689851

Wang, L., Lipsey, K., Murray, C., Prendergast, N., & Schoening, P. (2007). The Bioinformatics Program at Washington Universitys Bernard Becker Medical Library: Making it happen. *Medical Reference Services Quarterly*, *26*(2), 87–98. doi:10.1300/J115v26n02_08 PMID:17522011

Watson, H. J. (2014). Tutorial: Big Data Analytics: Concepts, Technologies, and Applications. *Communications of the Association for Information Systems*, *34*, 65.

Weiss, S., Indurkhya, N., Zhang, T., & Damerau, F. (2004). *Text Mining: Predictive Methods for Analyzing Unstructured Information*. New York: Springer-Verlag.

Wernicke, S., & Rasche, F. (2006). FANMOD: A tool for fast network motif detection. *Bioinformatics (Oxford, England)*, *22*(9), 1152–1153. doi:10.1093/bioinformatics/btl038 PMID:16455747

Wheeler, D. L., Barrett, T., Benson, D. A., Bryant, S. H., Canese, K., Chetvernin, V., & Federhen, S. et al. (2007). Database resources of the national center for biotechnology information. *Nucleic Acids Research*, *35*(suppl 1), D5–D12. doi:10.1093/nar/gkl1031 PMID:17170002

Wheeler, D. L., Church, D. M., Federhen, S., Lash, A. E., Madden, T. L., Pontius, J. U., & Wagner, L. et al. (2003). Database resources of the National Center for Biotechnology. *Nucleic Acids Research*, *31*(1), 28–33. doi:10.1093/nar/gkg033 PMID:12519941

Whittaker, R., & Smith, M. (2008). M-health – using mobile phones for healthy behavior change. *International Journal of Mobile Marketing*, *3*(2), 80–85.

Willinsky, J. (2006). *The access principle- the case for open access to research and scholarship*. Cambridge, MA: The MIT Press.

Wilson, P., & Risk, A. (2002). How to find the good and avoid the bad or ugly: A short guide to tools for rating quality of health information on the internet. *British Medical Journal*, *324*(7337), 598–602. doi:10.1136/bmj.324.7337.598 PMID:11884329

Wilson, T. D. (1981). On user studies and information needs. *Journal of Librarianship*, *37*(1), 3–15.

Wong, L. (2002). Datamining: Discovering Information from Bio-Data. In T. Jiang, Y. Xu, & M.Q. Zhang (Eds.), Current topics in computational molecular biology, (pp. 317-342). Cambridge, MA: MIT Press.

Wong, L. (2002). Technologies for Integrating Biological Data. *Briefings in Bioinformatics*, *3*(4), 389–404. doi:10.1093/bib/3.4.389 PMID:12511067

Worldometers. (2014). *Real time world statistics*. Retrieved from http://www.worldometers.info/world-population/

Wu, C., & Nebert, D. W. (2004). Update on genome completion and annotations: Protein Information Resource. *Human Genomics*, *1*(3), 229. doi:10.1186/1479-7364-1-3-229 PMID:15588483

Wu, M., & Chen, S. (2011). Graduate students usage of and attitudes towards e-books: Experiences from Taiwan. *Program*, *45*(3), 294–307. doi:10.1108/00330331111151601

Wu, W. G., & Li, J. (2007). RSS made easy: A basic guide for librarians. *Medical Reference Services Quarterly*, *26*(1), 37–50. doi:10.1300/J115v26n01_04 PMID:17210548

Wu, X., Kumar, V., Ross Quinlan, J., Ghosh, J., Yang, Q., Motoda, H., & Steinberg, D. et al. (2008). Top 10 algorithms in data mining. *Knowledge and Information Systems*, *14*(1), 1–37. doi:10.1007/s10115-007-0114-2

Xue, L., Yen, C. C., Chang, L., Chan, H. C., Tai, B. C., Tan, S. B., & Choolani, M. et al. (2012). An exploratory study of ageing womens perception on access to health informatics via a mobile phone-based intervention. *International Journal of Medical Informatics*, *81*(9), 637–648. doi:10.1016/j.ijmedinf.2012.04.008 PMID:22658778

Yadav & Tawmbing. (2015). *Use of bioinformatics resources & tools by users of bioinformatics centre in India*. Retrieved October 12, 2015, from http://digitalcommons.unl.edu/cgi/viewcontent

Yarfitz, S., & Ketchell, D. S. (2000). A library-based bioinformatics services program. *Bulletin of the Medical Library Association*, *88*(1), 36–48. PMID:10658962

Yi-Bu, C., Chattopadhyay, A., Bergen, P., Gadd, C., & Tannery, N. (2007). The Online Bioinformatics Resources Collection at the University of Pittsburgh Health Sciences Library System--a one-stop gateway to online bioinformatics databases and software tools. *Nucleic Acids Research*, *35*(1), D780–D785. PMID:17108360

Yi-Ping Phoebe, C. (Ed.). (2005). *Bioinformatics Technologies*. Berlin: Springer-Verlag.

Yogesh, K. S. (2006). *Fundamental of Research Methodology and Statistics*. New Delhi: New Age International Publishers.

Yooseph, S., Sutton, G., Rusch, D. B., Halpern, A. L., Williamson, S. J., Remington, K., & Li, W. et al. (2007). The Sorcerer II Global Ocean Sampling expedition: Expanding the universe of protein families. *PLoS Biology*, *5*(3), e16. doi:10.1371/journal.pbio.0050016 PMID:17355171

Youngkin, A. (2010). Librarian-controlled RSS: A novel approach to literature search follow-up. *Journal of Hospital Librarianship*, *10*(2), 123–131. doi:10.1080/15323261003680028

Zaharia, M. (2012). Resilient distributed datasets: A fault-tolerant abstraction for in-memory cluster computing. In *Proceedings of the 9th USENIX conference on Networked Systems Design and Implementation*. USENIX Association.

Zdobnov, E. M., Lopez, R., Apweiler, R., & Etzold, T. (2002). The EBI SRS server-new features. *Bioinformatics (Oxford, England)*, *18*(8), 1149–1150. doi:10.1093/bioinformatics/18.8.1149 PMID:12176845

Zhang, S. L. (2007). The flavors of open access. *OCLC Systems & Services: International Digital Library Perspectives*, *23*(3), 229 – 234.

Zhou, Y., & Ramona, B. (2011). Mobile options for online public access catalogs. *iConference '11 Proceedings of the 2011 iConference*. Retrieved October 8, 2011, from http://dl.acm.org/citation.cfm?id=1940842

About the Contributors

Shri Ram works at Thapar University. He worked as Deputy Librarian at Jaypee University of Information Technology, Solan since 2002. Earlier, he has worked at Indian Council of Medical Research, New Delhi and CEMCA India in various capacities. After completing a double Master's in Anthropology and Library and Information Science with NET, he was awarded his PhD degree in Library and Information Science from Osmania University, Hyderabad. Recently, he has been awarded with Commonwealth Professional Fellowship at University of East London, UK. He has published more than 45 papers in different peer reviewed international and national journals and conferences of international & national repute. He has edited three international conference volumes and attended more than 20 conferences of international and national repute.

* * *

Pavan Kumar Agrawal is an Associate Professor in Department of Biotechnology, GB Pant Engineering College, he holds a PhD. in Biotechnology from Barkatullah University Bhopal. He received research fellowship award from MOeF, and DBT, New Delhi during his research work. He is an active researcher engaged in the field of Microbial Technology and its application in bioinformatics for data retrieval. He has more than 30 research paper in various national and international journals to his credit.

Reysa C. Alenzuela is currently having her Post Doctoral Research at Kyungpook National University under National Institute for International Education Korean Government Scholarship Program. She finished her Doctor of Philosophy at the University of Iloilo, Master's degree in Library and Information Science from University of the Philippines – Diliman and Bachelor's degree in Secondary Education major in Library Science and English at University of San Agustin (Cum Laude). She is on Sabbatical leave as the Dean of Academic Affairs of Cabalum Western College and Chief Librarian of Iloilo Doctors College. She was a lecturer at Central

Philippine University and University of San Agustin. Before becoming a full time academician, she served as the Director of Thomas Jefferson Information Center (2007-2012) where she handled 47 networks of academic and public libraries all over the Philippines. Aside from her full time jobs, she is working to develop local consortia, conducting research in Archives, Library and Information Science curriculum development and disaster preparedness for libraries. She is also one of the pool of accreditors for Philippine Association of Colleges and Universities Commission on Accreditation overseeing the quality of academic libraries.

John Paul Anbu is currently the head of Periodicals at the University of Swaziland Libraries, Swaziland. His contributions include an open source circulation control module for UNESCO's CDS-ISIS, an information system on "Tea Disease Management" for United Planters Association of Southern India and a number of automation and digitizing projects in India and Swaziland. He has also contributed close to 45 research articles in various journals and books and co-edited 3 volumes of books. He is instrumental in piloting the SMS based content alert system for the University of Swaziland through the Emerald Publishers. He is a regular invited speaker in a number of International Conferences. His research paper "Changing Face of Libraries: Emerging New Technologies for Libraries" received the best paper award at ETTLIS2008 in New Delhi and "Libraries on the Move: Emerging Mobile Applications for Libraries" won the SATKAL Leading Edge Paper award in June 2010. He is the Swaziland's country coordinator for eIFL-FOSS and INASP.

Seetharaman Balaji received his Master of Science, Master of Philosophy and PhD in Bioinformatics. His doctoral research was on Biochemical informatics from Mangalore University. He has delivered several Guest Lectures on various aspects of Bioinformatics most notably, for the M.Tech Computer Science & Engineering students of NIT, Calicut. He is currently working as an Associate Professor in the Department of Biotechnology, Manipal Institute of Technology, Manipal University. He is a recipient of young scientist award for the year 2010-11 from the Govt. of Karnataka. He is also guiding 4 PhD scholars. He has published 2 monographs, 4 book chapters and over 30 research papers in International Journals.

Punit Gupta is Assistant Professor (Grade-I) in the Department of Information Technology, Jaypee University of Information Technology (JUIT), Waknaghat, Solan-173215, Himachal Pradesh, India. He received B.Tech. Degree in Computer Science and Engineering from Rajiv Gandhi Proudyogiki Vishwavidyalaya, Madhya Pradesh in 2010. He received M.Tech. Degree in Computer Science and Engineering from Jaypee Institute of Information Technology (Deemed university) in 2012 On

"Trust Management in Cloud computing". He is a Gold Medalist in M-Tech. He Joined Jaypee University of Information Technology Waknaghat on February 2013.

Ravi Jha, M.tech researcher student in Jaypee university of Information and Technology, Wakanaghat, Solan. He received M.Tech. Degree in Computer Science and Engineering from Jaypee university of Information and Technology, wakanaghat in 2016 On "Data Synchronization in Machine to Machine Communication through the cloud ". He received B.Tech. Degree in Computer Science and Engineering from Rajiv Gandhi Proudyogiki Vishwavidyalaya, Madhya Pradesh in 2010.

Viveka Vardhan Jumpala is serving as an Assistant Professor, Department of Library and Information Science, Osmania University (OU), Hyderabad. He obtained M.Tech., M.Sc. M.L.I.Sc, M.Ed from Osmania University. He was awarded Ph.D from Osmania University, Ph.D thesis entitled as 'A study of Web search Engines with Particular Reference to Information Retrieval Features'. He has qualified in the UGC-SLET in LIS in1997 and contributed about 30 articles to conference volumes and journals. He has authored a book entitled 'Manual of Library and Information Science' for UGC-NET/SLET. He is working as Counselor at B.R. Ambedkar Open University, Hyderabad. He is Author for State Government science text books, involved in various confidential activities. He is life member of IATLIS, APLA, ALSD, TSLA. Member BoS in OU. He is recently visited Thailand to present a seminar paper at Mahasarkham University.

Icxa Khandelwal has completed her degree in Bioinformatics from Jaypee University of Information Technology, Waknaght, Solan. She worked on various academic projects using various bioinformatics database resources. She created a sequence analyser tool utilizing sequences from NCBI database. As a part of her training at Jawaharlal Nehru University, Delhi, she worked on protein structures from PDB as well as inhibiting compounds from PubChem and other similar databases for analysis and identification of potential inhibitors. Her research experience includes Meta-analysis of Brain and CNS cancer microarray datasets, using gene expression data from Oncomine and other related information from GEODataset and Gene databases, to further identify the significant pathways using KEGG.

Sudhir Kumar is Former Professor & Head of School of Studies in Library & Information Science, Vikram University, Ujjain. He holds Master degrees in Arts and Library & information science. Has 33 years of teaching experience to Post Graduates, 1 year to M.Phil and 15 years experience in guiding the research scholars. 6 students have been awarded Ph.D. under his supervision. He has published 90 papers in journals, conferences and seminar volumes. He has received award of

Commendations for paper presentation in National Conference. He has two books to his credit. He is life member of various professional bodies. Currently, he is chairman of Board of Studies in Library & information science.

Manlunching was interest in literature and education. She completes her higher studies in university of Delhi, Delhi and awarded Ph.D in Library and Information Science in May 2015. Presently she is working in Saha Institute of Nuclear Physics, Library Section as Scientific Assistant - 'B'. She is confident and enthusiast in nature and she strive to be always the best in her future endeavor as well as in her work. Success is her ambition in her field of work. She never missed the good opportunity when fortune favors her.

Ntombikayise Nomsa Mathabela is an Assistant Librarian -Law at the University of Swaziland, in Southern Africa. She has worked as the Law librarian since 2010, where she also takes part in teaching library skills to undergraduate Law students, as well as school librarianship to post graduate students at the University of Swaziland. She is a former legal officer at the University of Swaziland where she was involved in legal affairs and personnel issues for the University of Swaziland staff and students. She holds a Bachelor of Laws from the University of Swaziland, a Post graduate Diploma from the University of Cape Town and a Masters in Library and Information Studies from the University of Alabama, Tuscaloosa - USA. She is currently pursuing her PhD in Library and Information Science at the University of KwaZulu-Natal. She is a 2009 Fulbright Scholar and an OCLC/IFLA 2014 Fellow. She is an active member and assistant pastor in her church and local community especially for Women and Youth ministry where she is involved in counselling and motivational activities. She has been the leader of the Women's wing for the past six years where she has successfully directed the compilation of the constitution and mode of operation as well as information dissemination to congregants. She is actively involved in the Swaziland Library and Information Association (SWALA) as secretary. Her interest is in building strong library associations and improving the library profession. She is the also member of International associations such as LIASA, ALA and ACLA. She can be e-mailed at nnmathabela@gmail.com / mathax@uniswa.sz

Shubhada Nagarkar worked in the Bioinformatics Center, Savitribai Phule Pune University, (former University of Pune) during December 1989 to December 2005 as Librarian / Sr. Tech. Assistant. Working as Assistant Professor, Department of Library and Information Science, Savitribai Phule Pune University (former University of Pune) since 2006 and teach 10 papers in four semesters every year. Designed and taught three new papers. Recognized Ph.D. and M.Phil. guide of Savitribai

Phule Pune University and S.N.D.T. University, Mumbai. Fulbright Scholar during 1998-1999 and visited to Mortenson Center and Graduate School of Library and Information Science, University of Illinois, Urabana-Champaign, USA. Recipient of International Travel grants by National Science Foundation, USA (2009), Ford Foundation, USA (2004), GBIF, Copehnhegen, (2003), OCLC, USA (2001) to participate and present papers in various international conferences and workshops. Recipient of Young Teacher Travel Grant (2014) and Unassigned Travel Grant (2001) of Savitribai Phule Pune University. Published 20 papers in various national and international conference proceedings and in journals. Completed three Research Projects of duration of two to three years. Currently working on the development of "Research Portal" of the Savitriabi Phule Pune University, Pune, India. Secretary, Pune University Library and Information Science Alumni Association (PULISAA) since last four years. More details at http://sites.google.com/site/pulisaa, Advisory Board member of Cameo Pvt. Ltd. Company, Pune.

Priya Panigrahi did her PhD in Bioinformatics from JUIT under the supervision of Dr. Tiratha Raj Singh.

Rekha Pareek completed her masters in Library and Information Science from Vardhman Mahaveer Open University, Kota and presently working for her Ph.D. as a research Scholar in University of Kota, Kota.

Manika Sehgal has completed her doctoral studies in Bioinformatics from Jaypee University of Information Technology. She has been working as Research Associate since then at ICAR- NBAGR, Karnal. Recently, she has joined Dr. Raghava's lab at CSIR- Imtech Chandigarh as DST- National Post Doctoral Fellow.

Aditi Sharma completed her degree in Bioinformatics from Jaypee University of Information Technology, Waknaghat, Solan. She has hands-on expertise of several bioinformatics databases and tools. She created an Enzyme database, for which she explored various existing enzyme databases such as Brenda, etc. Her research experience includes Meta-analysis cancer microarray datasets, using Oncomine, GEODataset and Gene databases, to further identify the significant pathways using KEGG.

Rahul Shrivastava is Assistant Professor (Senior Grade) and Coordinator of Training and Placement at Jaypee University of Information Technology, Solan (H.P.). He has served as Senior Research Associate at Evalueserve.com Pvt Ltd, a leading database and research analytics company. He has more than 15 years experience of handling biological data and resources during his PhD from Central Drug

Research Institute, a leading CSIR drug discovery institute. His research interests include bioinformatics analysis and usage of database and resources particularly in the area of microbial pathogenesis and drug discovery. He has many national and international publications to his credit. He has presented papers, oral technical session & posters at national and international conferences, and has been invited for guest lectures by Universities of repute.

Tiratha Raj Singh did his PhD in Bioinformatics from MANIT, Bhopal and joint research work for PhD was also done from Tel-Aviv University, Tel-Aviv, Israel. He worked with Dr. KR Pardasani and Dr. Tal Pupko; their role as supervisor and co-supervisor for my PhD. After completing his PhD in 2008, he won post doctoral fellowship from planning and budgeting committee (VATAT), TAU, Israel. He did his post doctoral studies in Molecular Evolution and Functional Genomics in Dr. Dorothee Huchon's lab. Returning from Israel, he joined DAVV, Indore, India for a year during 2009-10. He joined JUIT in July 2010. His research interests include various aspects of genomics, proteomics towards their involvement in various human diseases along with the study on annotation of complex Biological Networks through Computational Systems Biology approach.

Satyabati Devi Sorokhaibam is currently working as an Acting Librarian in the University of Swaziland, in Southern Africa. She carries with her a rich experience and expertise of having worked with the prestigious institute INFLIBNET (Information and Library Networked) Centre an IUC of UGC prior to joining this institute as a Project Scientist with the bulging project "UGC-INFONET Digital Library consortium" in India. She authored more than 60 articles/papers journals and seminar/conference proceedings in Regional, National and as well as in International level. She has also contributed papers for collected works in book chapters and festschrift volumes. She attended a number of seminars and conferences. She has worked on a project on Bibliography Compilation of Manipuri Literature under Sahitya Akademi, New Delhi. She is an International Advisory board member of TAV Dr. TAV Murthy Festschrift Volume, 2012-13 and foreign Advisory Board member of the Journals e-Library Science Research Journal (LSRJ). She also review the articles of the Journals "Issues in Business Management and Economics" and "Sky Journal of Education Research". She is an examiner of Ph. D. thesis of Bharathidasan University, Tiruchirappalli and Bharathiar University, Coimbatore. She also give lectures to Students of Manipur University Library Department and also in (Jawaharlal Nehru University) JNU lectures series on her visit to India. She attended a number of seminar and conference at Regional, National and International level. She is the member of the National and International associations including SIS, SWALA, MALA and LIASA She can be e-mailed at satyabati2005@gmail.com.

Index

F

G

H

I

J

K

L

M

www.ingramcontent.com/pod-product-compliance
Lightning Source LLC
LaVergne TN
LVHW062117200926
848484LV00007B/1699